Making
Sense

Making Sense

A Student's Guide to Research and Writing

Geography and Environmental Sciences

Fourth Edition

Margot Northey / David B. Knight /
Dianne Draper

OXFORD
UNIVERSITY PRESS

OXFORD
UNIVERSITY PRESS

8 Sampson Mews, Suite 204, Don Mills ON, M3C 0H5
www.oupcanada.com

Oxford University Press is a department of the University of Oxford.
It furthers the University's objective of excellence in research, scholarship,
and education by publishing worldwide in

Oxford New York

Auckland Cape Town Dar es Salaam Hong Kong Karachi
Kuala Lumpur Madrid Melbourne Mexico City Nairobi
New Delhi Shanghai Taipei Toronto

With offices in

Argentina Austria Brazil Chile Czech Republic France Greece
Guatemala Hungary Italy Japan Poland Portugal Singapore
South Korea Switzerland Thailand Turkey Ukraine Vietnam

Oxford is a trade mark of Oxford University Press
in the UK and in certain other countries

Published in Canada
by Oxford University Press

Library and Archives Canada Cataloguing in Publication

Northey, Margot, 1940-
Making sense: a student's guide to research and writing in geography and
environmental sciences / Margot Northey and
Dianne Draper. — 4th ed., rev. with up-to-date MLA & APA

Includes bibliographical references and index.
ISBN 978-0-19-544002-7

1. Geography—Authorship.   2. Environmental sciences—
Authorship.   3. English language—Rhetoric.   4. Report writing.
I. Draper, Dianne Louise, 1949–   II. Title.

G74.N67 2010   910.01'4   C2010-900137-0

Cover image: mariusFM77 / iStock Photography

5 6 7 – 12 11 10
This book is printed on permanent (acid-free) paper ∞.
Printed in Canada

Mixed Sources
Product group from well-managed
forests, and other controlled sources
www.fsc.org Cert no. SW-COC-002358
© 1996 Forest Stewardship Council
FSC

Table of Contents

Preface

The first edition of this book, published in 1992, with its numerous suggestions to help the college or university student, clearly addressed a need; it was widely adopted as a supplementary text for students in geography, the environmental sciences, interdisciplinary environmental studies, and other disciplines, including history. The second edition, published in 2001, incorporated changes that had resulted from the remarkable developments in computer technology, and, among many other things, responded to student suggestions for a reordering of the chapters. The innovative integration of reference styles proved to be very useful. The appendix on weights, measures, and notation was added, and is a readily available and valuable resource for users of the book. An updated second edition, published in 2005, included the newly made revisions to CBE, APA, and MLA referencing guidelines.

The third edition, published in 2007, responded to the continuing computerization of libraries and to the many research possibilities available on the Internet. New expanded and rewritten sections were provided regarding source materials available in the library, computerized library catalogues, and research material available on the Internet, including an updated list of major Internet sites of special merit for students in geography, the environmental sciences, and environmental studies. The section on plagiarism and cheating was clarified, and the section on research ethics was expanded. New materials were added to the chapter on writing, reading, and lecture notes, to the sections addressing take-home exams and preliminary material in a thesis, to the use of visual aids, and to the section on computer-generated digital mapping. Additional material was added to the sections on annotated bibliographies, preparing a questionnaire, formatting and protecting your work, and GIS-based projects published online. The core chapters on different types of writing, style, use of words, common errors and usage, punctuation, documentation, and definitions remained, as did the catch list of misused words and phrases, each with some changes related to the above material.

Many of the revisions in this fourth edition are due to continuing technological developments and the ongoing concern for research ethics that affect how research is conducted and reported, the learning process, and the way we write. In addition to many small changes we have made, larger changes to this edition include the following:

- an expanded discussion of Internet sources (and pitfalls in their use) includes new material to help you assess the validity of different types of websites as academic research resources;

- a substantial section on group work has been added;
- references have been revised and updated;
- a new figure has been added to clearly identify Sheridan Baker's funnel/inverse funnel approaches to writing; and
- a new diagram on preparing a proposal has been included.

In addition, the list of journals that are of special interest to geographers and environmental scientists has been expanded and placed in a new appendix for easier reference. Likewise, the list of pertinent websites has been expanded, and it, too, has been relocated to an appendix.

Acknowledgements

The fourth edition of *Making Sense in Geography & Environmental Sciences* builds on the excellent work of David Knight and Margot Northey. For this new edition a co-author, Dianne Draper, was invited to join the team to undertake the revisions. Having enthusiastically recommended the text to students since the first edition was published, she was pleased to contribute her expertise to the endeavour of helping students improve their academic performance.

The authors are grateful to the many students and faculty who over the years have made suggestions for improving the book. The people who assisted David Knight in the preparation of the first three editions were acknowledged in those volumes. Dianne Draper gratefully acknowledges Marilyn Kinnear and Emma Stewart, who kindly identified some references to cite. As before, the illustrations were expertly drafted by Christine Earl (Figure 8.1), Paxton Roberts (Figure 8.2), Stefan Palko (Figures 8.3 and 8.4), and Marie Puddister (7.3, and 8.5 to 8.15).

In addition, the authors gratefully acknowledge the benefits derived from their editorial responsibilities with publishers and journals: David Knight as an editor with the *Encyclopaedia Britannica* and the Carleton University Press, and as a member of the editorial boards of several journals, including *GeoJournal*, *Journal of Cultural Geography*, and *Political Geography*; Dianne Draper as a member of the editorial board of the University of Calgary Press and *The Canadian Geographer*.

The authors are grateful for the expert assistance of several fine people at Oxford University Press, including Phyllis Wilson, Euan White, Ryan Chynces, Peter Chambers, Eric Sinkins, and Janice Evans.

Finally, thanks to the students who will use this text. May the contents of this book help contribute to your scholarly competence and performance.

David B. Knight   Dianne Draper
Elora, Ontario   Calgary, Alberta

A Note to the Student

This book is intended to teach students in geography, other environmental sciences, and interdisciplinary environmental studies programs how to undertake and successfully complete assignments that involve research and writing.

As the second chapter in this book notes, research takes many forms. How you report your research will vary from assignment to assignment. This book presents, as simply as possible, comments and suggestions to help you tackle the different types of work that you are expected to do. While the principal goal is to assist you in reporting on your reading, thinking, and research so that others may learn from you, you also may find that these comments are helpful as you read and listen and seek to make sense of others' research.

On writing assignments

Contrary to many students' belief, good writing does not come naturally; even for the best writers it is mostly hard work, following the old formula of 10 per cent inspiration and 90 per cent perspiration.

Writing in university or college is not fundamentally different from writing elsewhere. Yet each piece of writing has its own special purposes, and these are what determine its shape and tone. *Making Sense in Geography and the Environmental Sciences* examines both the general precepts for effective writing and the special requirements of reporting the research process and findings in the social, physical, and biological sciences, for students in geography or certain interdisciplinary portions of other subjects that in some way deal with the environment. It also points out some of the most common errors in student composition and suggests how to avoid or correct them. Written mostly in the form of guidelines rather than strict rules—since few rules are inviolable—this book should help you escape the common pitfalls of student writing and develop confidence through an understanding of basic principles and a mastery of sound techniques.

Organization

The organization of this book reflects linkages between clusters of chapters, as follows:

- *Chapters 1 and 2* deal with fundamentals of research and writing. Among the topics addressed in Chapter 1 are initial thinking and ways to approach a writing assignment, sources, how to start writing, basic evaluation criteria, plagiarism, and writing with a word processor. Chapter 2 begins with a discussion of the nature of research, the sources of human knowledge, the spectrum of approaches to doing research, ethical issues, and the formation of hypotheses; it then offers pointers that will help you to do research both in libraries and with online computer sources.

- *Chapters 3 to 6* focus on four different types of assignments: taking reading and lecture notes (which will form the foundation of your studies); writing a report on a book or an article; researching and writing an essay; and writing a lab report.
- *Chapter 7* outlines ways of preparing to be a full participant in discussion groups, seminars, tutorials, and conferences, suggests an approach to organizing and managing group work, and discusses the use of various visual aids—including photographs and posters—in presentations.
- *Chapter 8* focuses on illustrative material: maps, tables, equations, and graphs.
- *Chapters 9 and 10* offer suggestions concerning field work (preparing for, taking field notes, writing a field report) and major research papers and theses (writing a proposal, organizational patterns, basic structures).
- *Chapter 11* identifies how to include quotations in your written work and explains several documentation systems appropriate to different types of assignments.
- *Chapter 12* considers examinations (preparing for and writing them).
- *Chapters 13 to 18* focus on the fundamentals of writing, starting with a discussion of the importance of words and then considering basic issues: style in writing, common errors in grammar and usage, punctuation, frequently misused words and phrases, and definitions.
- Useful information on weights, measures, and notation can be found in *Appendix I* beginning on p. 300, while *Appendix II* contains a selected list of journals that are of interest to geographers and other environmental scientists. *Appendix III* provides a list of websites of special interest to geographers and others concerned with environmental issues. Finally, the *Index* beginning on p. 311 will help you to find the specific information you want.

CHAPTER 1

Writing and Thinking

Thinking and writing are the most basic skills you need to successfully complete your college or university education. As you begin to use this book, consider your role in the learning process, as outlined below.

- The instructor *offers* (readings, lectures, seminars, discussion groups, lab and field work assignments, etc.), *shares* (by imparting new knowledge and identifying research methods), *provokes* (by raising questions and providing insights), *responds* (to questions), and *evaluates* (assignments and examinations).
- You *listen* (to lectures), *participate* (in class interaction), *read* (assigned and additional readings), and *do* (lab, field work, and other assignments).
- You, the learner, *take in*, *think about*, *mull over*, *question*, *analyze*, *evaluate*, *integrate*, *"report back"* (via essays, reports, and examinations), and *"take away"* (all that you gain from a course and, ultimately, all of your courses).

The key message here is that you need to take an active role in the learning process. *You* are the key to *your* education! This book offers many suggestions that will help you make sense of what you are asked to do and will enhance your skills as you seek to do your best with different types of assignments in your courses, whether they are in physical and human geography, agriculture, biology, geology, landscape architecture, environmental history, other environmental sciences, or interdisciplinary environmental studies. To start, this chapter highlights writing and thinking.

Writing is a process of many stages, every one of which involves thinking. You are not likely to produce clear writing in the final copy of your assignment unless you have first done some clear thinking, and clear thinking can't be hurried. It follows that the most important step you can take is to leave yourself enough time to think.

Among the other steps that the writing process requires—long before you complete your final paper for submission—are selecting the topic, framing your work within the context of the discipline and course for which it is intended, developing a thesis statement or examining a hypothesis, drafting, revising, and editing. This chapter discusses all these aspects of the writing process, and concludes with some comments on evaluation criteria, plagiarism, and the use of a computer for preparing assignments.

USING YOUR SUBCONSCIOUS

It is useful to be aware, at the outset, of a problem that eventually puzzles all writers, no matter how skilled: that infamous writer's block. You know the symptoms: "What am I going to write about?" "How am I going to write about this?" "How can I write convincingly when I am unsure about the purpose of the essay?" "How do I define what it is I plan to do?" "What is my key question?" "Do I need a hypothesis, or is a simple statement of purpose enough?" "What can I do to make sense of this material?" "Should I use subheadings?" "How am I ever going to come to a conclusion?" A good thing to remember is that we do not always solve a difficult problem by "putting our mind to it"— by determined reasoning. Sometimes it's best to take a break, turn to something else, perhaps even sleep on it, and let the subconscious or creative part of your brain take over for a while. Very often a period of relaxation will produce a new approach or solution. Leaving time for creative reflection, of course, does not mean sitting around listening to CDs or watching TV until inspiration strikes out of the blue. Insights that otherwise might escape you may come if you set aside some quiet moments for reflection, away from your books; try going outside and thinking about the assignment on and off as you walk, run, or bicycle and look at the world around you. Invest in yourself! The dividends can be considerable.

GUIDELINES FOR ANY WRITTEN WORK

With almost everything you write there are some basic guidelines that you should keep in mind:

- Think about the purpose and the context of the work you have to produce.
- State clearly what you plan to do.
- Define your terms.
- Include only relevant material.
- Strive for consistency of expression throughout the work.

- Make sure you are accurate in all of your statements, including your analysis and presentation of data.
- Present your information in a logical and effective order.
- Convey your message as simply and clearly as possible.
- Make sure that your work is both coherent and complete.
- Do not draw conclusions that are not clearly based on your evidence.
- Never assume that one draft will do the job; count on writing at least two drafts before producing the final copy.
- Provide documentation for your sources and ensure that your references are complete and accurate.
- Be consistent in style throughout the assignment.
- Attend to the small things: give the work a title, select the most appropriate format for subheadings, and number the pages.
- *Always* proofread and make any necessary corrections before submitting the work.
- Remember to include your name and student number, the name of the instructor, the course number, and the (due) date.

INITIAL STRATEGIES

To write is to make choices. Practice makes the decisions easier to come by, but no matter how skilled you become, with each piece of writing you still will have to choose. Even if the instructor indicates what you must write about, you still must take the time to reflect and make choices.

You can narrow the field of choice from the start if you realize that you are *not* writing for anybody, anywhere, for no particular reason. At school (or anywhere else), it is always sound strategy to ask yourself four basic questions:

- "What is the purpose of this piece of writing?"
- "What is the reader like?" Or, to put it other ways, "Who is my audience?" or "Who am I writing for?"
- "What does the reader expect?"
- "What do I expect of myself?"

Your first reaction may be, "Well, I'm writing for my instructor to satisfy a course requirement," but obviously that's an inadequate response. To be useful, your answers have to be precise. Thus, a better answer might be, "I am going to work with data *X* using methodology *Y* so that I can then write a research report for course Z." And keep in mind your personal investment in the assignment: presumably you will always want to do the best work you can.

Think about the purpose

Your purpose may be any one or more of several possibilities:

- to show that you understand certain terms, concepts, or theories
- to establish that you can do independent research
- to indicate that you can apply a particular research method
- to report on a test you have performed
- to apply a specific theory to new material
- to provide information
- to reveal your knowledge of a topic or text
- to present data you have gathered while doing field work
- to demonstrate your ability to evaluate secondary sources
- to illustrate your creativity
- to show that you can think critically

Certainly, an assignment designed to see if you have read and understood specific material calls for a different approach from one that is meant to test your critical thinking or research skills. If you don't determine the exact purpose, you may find yourself working at cross-purposes—and wasting a lot of time.

Think about your approach

Tied to the purpose for any particular assignment is the issue of approach. Doing research means that you have to make choices before you start; different approaches will lead you to pose different types of questions and to use different methods. Your approach to any piece of writing depends in part on what course the assignment is for, since there are disciplinary (and subdisciplinary) differences in approaches to research and writing. How did the instructor describe the purpose of the course at the beginning of classes? What are the course's essential concepts and presuppositions? How does this particular assignment relate to the latter? What are the key themes, processes, and methods that you have heard about in lectures, discussion groups, and labs, or read about in assignments? How can you bring these various elements together to write something (be it an essay, field report, lab report, or whatever else) that is germane to the course? The links between research and writing are discussed in the following chapters.

Think about your thesis and hypothesis

For any research or writing, it is essential to have a clear purpose: in short, to know what it is that you are proposing to do (see Chapter 2). Isolate the key issue and write it down as your statement of purpose. Take the time to work out your purpose statement carefully. You might start by thinking about the key question or set of questions that you want to answer. Then, turn the key question into a statement. Use a complete sentence to express it, and above all make sure that it is *limited* (since you are not going to study everything), *unified* (so that there is a sense of wholeness to what you propose to do), and *exact* (since both you and your reader must be sure about what you intend to do).

Many types of research will require that you formulate a hypothesis. A hypothesis, or supposition, is a working thesis, an intended line of argument that you are free to change at any stage of your planning—though once you have actually started your research, you should stick to the hypothesis you have chosen. A hypothesis works as a linchpin, holding together your approach to the project, data, and ideas as you organize and do your research. A hypothesis helps you define your intentions and make your research selective. And, as a result of your exploration of a certain set of data, by the time you start to write up your report or essay you will have determined whether or not your working thesis was valid.

Think about your sources

Whatever you write will be based on some kind of research. Doing research and writing about it is not like writing a short story or novel based on personal reflection and imagination: as a researcher, you must get to know and understand your sources.

As you work your way through college or university you will become well acquainted with three kinds of sources. The most important *primary source* for all students of geography, environmental studies, and the environmental sciences is the earth around us and the people who inhabit it. In your courses you will learn how to "read" and understand the form and content of the earth's surface as modified by natural and human processes, and the ways people perceive, organize, and use their surroundings. Other *primary sources* include direct field observations, the materials used in labs for testing, questionnaire data, censuses, maps, land records, air photos, satellite images, and the like. Some books also can be primary sources, if they provide direct evidence that is of use to you. Primary sources vary from one discipline (and subdiscipline) to another; thus one type of study may rely on temperature readings as a source, while another may demand the use of land registry data, or fishing licence records, or tree rings, or herbs, or dolphins, or fertilizers, or water flows, or old newspapers, or cemetery stones, or barns.

Secondary sources, by contrast, are books and articles written by authors who have read and reflected upon a specific literature in order to provide a context for their research or to serve as the basis for certain generalizations. Since some of your research may well be based only on secondary sources, it's a good idea to visit the library regularly, so that you become familiar with the books and journals that relate to your interests. Some secondary sources also are available to you on the Internet (see Chapter 2).

Tertiary sources provide extended syntheses of previous research and generally do not have as their primary purpose the reporting of new research by the authors; most textbooks fall into this category.

Suggestions for finding library and other sources can be found in Chapter 2. Ways of reporting your findings are discussed in various other chapters, including (for a book report) Chapter 4, (for an essay) Chapter 5, and (for field work) Chapter 9.

Finally, unless your instructor tells you otherwise for a specific assignment, do not rely only on Internet-based searches. It is not enough—indeed, it is unwise—to examine only online material. Use the materials available to you on the shelves in libraries!

Think about illustrations

In many cases some of the material you will use can be mapped, graphed, or tabularized. Think carefully about the appropriateness of the material you have gathered and ask yourself the following questions:

- Is the data really necessary to support or expand upon a point or argument presented in your text?
- Would it be more useful to place the data in the text, or in a table, or in mapped or graphic form?
- If you decide to use an illustration, what is the most effective form for displaying the data? (see Chapter 8 for some suggestions).

Illustrations in academic work are not simply nice appendages that may (or may not) look good; they are an integral part of any work. As you write, then, you also must take into account the material you can show in graphic form. Use tables, graphs, maps, photographs, and other illustrations when they clearly and effectively will provide the reader with data that are essential to your case. And, of course, as you edit your work, you will need to ensure that you have correctly numbered and titled your illustrations and referred to them appropriately (by their numbers) in your text. In some circumstances, you may be able to display your illustrations on a poster (see Chapter 7).

Think about the reader

Thinking about the reader does not mean playing up to the instructor. To convince your reader that your own views are sound, you have to consider his or her way of thinking. If you are writing a paper on the environmental history of a particular place for a geography professor, your analysis and presentation obviously will be different from what they would be if you were writing for a professor in resource economics or landscape architecture. You will need to make specific decisions about the basic questions to be asked, the terms to be explained, the background information to be provided, the data to be used, and the details necessary to convince that particular reader that your discussion has merit and is valid. Remember that your reader will approach your work with certain assumptions and questions posed by his or her discipline and circumscribed by the nature and purpose of the course you are taking, so be sure to frame your answer accordingly. Try to anticipate questions that might be raised and seek to answer them. If you don't know who will be reading your paper—your professor, a teaching assistant (TA) who may be a discussion-group leader or lab assistant, or a graduate student marker—just imagine someone intelligent, knowledgeable, and interested, skeptical enough to question your ideas but flexible enough to adopt them if your evidence is convincing.

Think about disciplinary "languages"

Each discipline has its own basic "language." Although many disciplines share some aspects of their language, there usually are some terms or usages that are unique (on jargon, see p. 216). These "languages" will become evident as you listen to lectures and read course texts. To understand some words, you likely will need to look up definitions. But many terms you will hear and read are not covered adequately in common dictionaries—if they are covered at all. Hence, you will need to use disciplinary dictionaries. Many pertinent paperback dictionaries are available for geographers and other students of the environment. In addition, a comprehensive, up-to-date atlas (print or electronic) is a useful, even essential, study aid.

Think about length and value

Before you start writing, you will need to think about the length of your assignment in relation to the time you have available to spend on it. If both the topic and the length are prescribed, it should be fairly easy for you to assess the level of detail required and the amount of research you need to do. If only the length is prescribed, that restriction will help you decide how broad or how narrow a topic you should choose. You also should think about how important the assignment is in relation to the rest of your work for the course. A piece of work worth 10 per cent of your final grade will not demand as much attention as one worth 30 per cent.

STARTING TO WRITE

Once you have completed the necessary research, you will face what may be the hardest part of any assignment: actually starting to write. Consider the following methods; one of them may help you structure your thoughts.

- Write an opening statement that clearly indicates the purpose of the piece of work. This will help you to focus your thoughts.
- Jot down your title for the assignment and, if possible, a few subheadings; then start to fill in the sections one by one.
- List your thoughts in point form, in random order, as they come to mind. Then look to see if you can reorder the points so that they outline a logical development to your argument. If you are using a pen and paper, circling key points and linking them with lines to other points may assist you in ordering things. If you are using a computer you will be able to make use of a wide array of formatting options, including coloured text, highlighting, cutting, copying, and pasting. This step may be necessary before you can identify all the subheadings you will need.
- If you become frustrated with one section, turn to another. By the time you have completed it, ideas on the skipped section may come more easily.

Think about tone

In everyday writing you probably take a casual tone, but academic writing usually is more formal. The exact degree of formality required depends on the assignment and instructions you have been given. In some cases—say, if your cultural geography professor asks you to express yourself freely and personally in a journal—you may well be able to use an informal tone. Essays, reports, and theses, however, require a formal tone. What kind of style is too informal for most academic work? Here are the main signs:

USE OF SLANG

Although the occasional slang word or phrase may be useful for special effect, use of slang generally is not acceptable. The reason is that slang expressions usually are regional and short-lived: they may mean different things to different groups at different times. (Just think of how widely the meanings of *hot* and *cool* can vary, depending on the circumstances.)

THE RIGHT WORDS?

Be careful in your choice of words; for more detail, see Chapter 13. If the words you want to use may be hurtful to others, don't use them.

EXCESSIVE USE OF FIRST-PERSON PRONOUNS

A formal essay is not a personal outpouring, as in a personal letter to someone close to you; this means you should keep it from becoming I-centred. It's acceptable to use the occasional first-person pronoun if the assignment calls for your opinions—as long as they are backed by evidence. But you should avoid the *I think* or *in my view* approach when the fact or argument speaks for itself. Still, if the choice is between using *I* and creating a tangle of passive constructions, it is almost always better to choose *I*. (A hint: when you do use *I*, it will be less noticeable if you place it in the middle of the sentence rather than at the beginning.)

USE OF CONTRACTIONS

Contractions such as *can't* and *isn't* are not suitable for academic writing (except within a quotation), although they may be fine for informal writing (e.g., letters to friends, or this handbook). The problem with trying to avoid excessive informality is that you may be tempted to go to the other extreme. If your writing sounds stiff or pompous, you may be using too many high-flown phrases, long words, or passive constructions (see Chapter 14). When in doubt, remember that a more formal style will always be acceptable.

THE EDITING STAGE

Specific types of writing are addressed in later chapters. With everything you write, however, remember that editing your work is a critical stage. It cannot be emphasized enough that, whatever you are writing, you should never accept your first draft as the final product. In fact, your most serious writing may not begin until after you've written your first draft. Read what you have written to identify illogically developed sections, awkward phrasing, incomplete sentences, skimpy arguments, missing references, and questionable conclusions; you may even need to do more research. Once you have redrafted the material, read it again with a critical eye. Don't hesitate to slash unnecessary verbiage, alter sentences, or restructure paragraphs—and when you have done that, check your references again, for facts, style, and completeness. Two, three, or even more drafts may be required before you arrive at the final copy. In short, you must be not only a writer but an editor.

Writing and editing go together, but they are separate processes. In the writing process you will craft something into being; in the editing process you will critically assess and re-craft that work to make it better. A word of caution: as you prepare successive drafts of your work, be sure to number each one (draft 1, draft 2, and so on—both on computer files and on printed copies) so that you will not mix them up and accidentally submit an early draft instead of the final version. Finally, you can help yourself at the editing stage if you number the pages of each draft.

A few words about appearance

We've all been told not to judge a book by its cover, but the very warning suggests that we have a natural tendency to do so. Readers of essays find the same thing. A well-typed, visually appealing piece of work creates a receptive reader, and, fairly or unfairly, often earns a higher mark. If your instructor provides specific instructions, be sure to follow them, but if not, double-space the lines and leave wide margins on the sides, top, and bottom, framing the script in white. Leave three centimetres, at least, at the sides and top and four centimetres at the bottom, so that the reader has ample space to write comments. Number each page and include the title, your name and student number, the name of the instructor, and the course title and number. The appearance of the paper may be improved by using a staple at the top left corner, slanted so that the folded-over pages will not tear. Good looks won't substitute for good thinking, but they certainly will enhance it.

Evaluation criteria

Remember that your work will be evaluated, not literally weighed! Depending on the assignment, your reader will want to see whether you have arrived at the correct answer, or argued your case well. When reading an essay, he or she will want to be challenged by what and how you write. Reactions will depend to a large degree on how well you write. Some work will simply be "right" or "wrong," but other types of writing will evoke varying responses: "Nice reasoning"; "Interesting idea"; "Good point"; "A bit of a bore"; "Wow, this is heavy going"; "You sure know how to waffle"; "Dreadful!" After some effort, you should be able to get the first two or three of these responses to anything you submit—and none of the rest.

Most instructors will look for and consider the following points when evaluating your work:

- a clear statement of purpose (or thesis statement; see pp. 28–9)
- placement of work within a context provided by a review of pertinent literature
- quality and relevance of evidence
- accuracy
- completeness
- imaginative development (especially in an essay)
- logical organization
- clarity of expression
- correctness of grammar, spelling, and punctuation
- pertinence and quality of illustrations (maps, photos, tables, etc.)
- conclusion
- appropriate and accurate citation of sources
- references/bibliography (which you may have to annotate)

A warning about plagiarism and cheating

Plagiarism is a form of stealing. The intention is to cheat by submitting someone else's material as one's own. Other types of cheating also are deemed to be academic offences, such as submitting the same work to different instructors without their prior knowledge and joint consent. Within post-secondary institutions, penalties for *academic misconduct* range from a zero for the work in question to outright expulsion. You should read and take to heart the entry on academic misconduct in your academic calendar or other official documents listing regulations.

You commit plagiarism or otherwise cheat if you commit any of the following actions:

- submit the work of another person, in whole or in part, as your original work, whether purchased or borrowed

- fail to identify a quotation from someone else's work
- fail to document the source of many types of material—for exceptions, see below—that you have incorporated into your work
- fraudulently manipulate or submit false lab work, field work observations, or research data
- copy someone else's answer to a test, or lab or field study, and pass it off as your own

Of these issues, the focus here is on being sure that you fairly and correctly identify when you are using someone else's work within your own.

The way to avoid plagiarism is to give credit where credit is due. If you use someone else's idea, acknowledge it, even if you have changed the wording or just summarized the main points.

- If you quote something by copying an author's exact words, use quotation marks and indicate the source.
- If you paraphrase or summarize someone's material, identify the author and indicate the source.
- Use identifiers within your text when you present someone's ideas, observations, or conclusions (e.g., "Best asserts that . . ." or "As Woodward observed . . .") and include the appropriate identification of the source.
- Authoritative or distinctive material should be cited with the source duly noted, whether you use it to argue against, or perhaps to bolster, your own stance.
- Identify sources for anything that is not common knowledge; however, things like important dates (of, say, World War II or the founding of a country), the present population of the country, or the national anthem need not be documented. Furthermore, you don't have to cite sources to support facts such as, "the world is round," or that "Brian Berry and Audrey Kobayashi are prominent geographers" (though reference to some of their works might be appropriate).
- Do document unfamiliar facts or claims—statistical or otherwise— or any that are open to question.
- Remember, too, that information you find on the Internet must be properly acknowledged. Even though websites are accessible instantly, they are the property of the individual or organization that publishes them. Accordingly, give appropriate citation within your

work when you quote from or otherwise refer to such material (see Chapter 11).

Be sure to record in your reading notes when source material you have copied has been directly quoted or summarized (see Chapter 3). As you assemble your work, be careful as you take things (a fact, a sentence, a paragraph, your paraphrase) to include in your work so that you don't inadvertently copy something that is not original to you. As you write, constantly ask yourself if you are using your own words to present your own ideas. If not, record the source in the pertinent location within the body of your work and in the references, using the method appropriate to the course and assignment (again, see Chapter 11).

Don't be afraid that your work will seem weaker if you acknowledge the ideas of others. On the contrary, your work will be all the more convincing: serious academic treatises are almost always built on the work of preceding scholars, and credit is duly given to the earlier work.

WORKING IN A GROUP

You may be expected to do some of your work as part of a group. Teamwork is common in the work world, so this experience is valuable. Group interaction can make such assignments both easier and more fruitful than solo work, since each person will have a different perspective, set of skills, and approach to problem-solving. To be effective, your group must identify clearly the objectives of any assignment and divide the tasks (experimentation, library research, writing) equitably. It is imperative that the group reaches a consensus at all stages of the assignment, from problem identification to final word-processing (see pp. 103–6, 107). The successful group report is one in which all participants have had an equal voice.

Be aware of how grades are to be determined. Some instructors assign grades on a group basis, but others may not. Check to see if there is to be a group grade (determined by an overall evaluation of the submitted work) or an individual grade (based on the specific section or sections you worked on by yourself). If a group grade is to be assigned, all participants—from the eager, dedicated ones like yourself to the lazy, "let others do the work for me" variety—will end up with the same mark. In that case, all of you will want to ensure that everyone in your group is a full participant, and that all members do their best. It may help if you insist that all group members sign a contract, as a way of holding each person accountable to the group, and be sure to hold regular group meetings so each person can provide their updates to the entire group.

WRITING WITH A COMPUTER

The advent of the computer and relatively inexpensive, easy-to-use word-processing packages has made a tremendous difference in the way people write. The word processor can be a wonderful tool to assist you in your writing—if you use it judiciously.

Word-processing programs permit you to type and correct mistakes before they arrive on paper, rearrange your material for ease of reading, and print out a neat and tidy final copy. Most systems are easy to learn, and they can speed up your writing considerably. To some extent they can help to improve your writing skills, because they make it easier for you to revise—to add and delete material, correct information, and move paragraphs around. And most programs have the capacity to check spelling and grammar. Here are a few simple suggestions you might find useful in word-processing.

Type your material directly into the computer

Traditionally, people wrote out their papers by hand before they typed them, but you may feel comfortable typing your paper directly into the computer. Even if you don't type very well, it doesn't really matter; when you are writing a paper, the time spent thinking will far outweigh the time it takes to enter the words into the computer. A common argument against writing in this fashion is "I can't think and type at the same time." But once you try, you're almost certain to find it rewarding. Seeing your thoughts appearing in a legible form will help to keep you going. However, be sure to do this writing by referring to your handwritten organizational notes. Don't rely solely on your memory.

Try different ways of organizing your paper

Perhaps the most useful aspect of a word-processing system is that it allows you to move blocks of text around so that you can try out different structures for your paper. If you have ever reached the point where your handwritten version has become so complicated that you have to write the whole thing out again in order to make any sense of it, you will appreciate being able to make on-screen corrections and rearrangements. You can set up a new organizational structure and if, after reading it through, you don't like it, you can always go back to your original version.

Don't let the system rule your thinking

Seeing something neatly typed on a screen or on paper makes it seem more acceptable than messy handwriting, even though the quality of the work may be no different. Don't be fooled into thinking that quality typing replaces quality thinking. Read over your work with a critical eye, knowing that you can easily change something that is unsatisfactory. Remember that the word processor is a tool for you to use—no more than that.

Save regularly and back up your files

If you use a computer regularly, you know the importance of this advice; if not, take it to heart. There is nothing more agonizing than discovering that something has gone wrong and caused you to lose everything you had been working on. It doesn't happen very often, but everyone experiences it at least once, and always unexpectedly. There is one easy way around the problem: when you are writing, save your work file every two to five minutes. You can do this manually or have the computer set to do it automatically. Then, when a friend pulls the plug on the computer to turn on the TV, the most you will lose is the typing you have done since you last saved your file.

Don't keep your critical files only on the computer's hard drive; copy them to an external memory device, such as a CD, a DVD, or a portable USB device. Just be careful where you put your back-up files: you don't want to learn the hard way what happens when the CD that contains your only copy is lustily chewed by your dog, or when you spill a cup of coffee on it, or when you accidentally step on your USB stick and break the connecting end! Again, a little foresight will save you from losing everything. As soon as you complete a section of work, make an external back-up copy and keep it well away from the computer. Regular use of a back-up system could spare you a lot of frustration. In addition, be sure to give a title and number to each version of your work, in case you need to refer back to an earlier version. Don't conflate or otherwise merge versions unless you are absolutely sure you want to do so.

Reading and proofreading

Reading a lengthy paper section by section on a screen makes it difficult to gain a sense of the whole. It's better to print a hard copy and read and edit your work in that form before making any needed corrections on the machine. Be sure to proofread for overall comprehension as well as for errors in grammar and spelling. The computer's grammar checker can help you by raising questions about how you have written a sentence. But, be wary of its suggestions; they may be nonsensical in the context of your writing. The gram-

mar checker may catch such things as a missing comma, but other suggestions, if accepted, could alter your meaning. So, don't accept the grammar
checker's suggestions without thinking carefully about what *your* intention is.
Because of this concern, it is highly recommended that you not set the grammar checker to make changes to your text automatically. Also, although most
word-processing software lets you check spelling, sometimes "automatically"
as you type, software can't catch a word that is correctly spelled but incorrectly
used (for example, your spell checker will not catch that you have typed "kid"
instead of "kind"), so don't rely on it alone. When using the spell checker on
your computer, be sure to turn on the dictionary language setting you want
(e.g., Canadian-English, English-US, English-UK, or Australian-English). In
fact, "correct" spelling itself can be an issue; see the next section.

Using dictionaries and thesauruses

A good dictionary is a wise investment; get into the habit of using one. It will
give you not only common meanings but less familiar applications, archaic
uses, and derivations, as well as "correct" spelling. The reason "correct" is in
quotation marks is that, to a degree, spelling varies. American English has different spelling and usage from British English, and Canadian spelling and
usage follow either British or American practices, but usually combine aspects
of both. Before you buy a dictionary check to make sure that it gives these
variants.[1]

Be aware that most print dictionaries, and the dictionary found in your
computer's word processing program, are not sensitive to disciplinary definitions. There are many disciplinary dictionaries.[2] Words in them may be
defined in ways that differ from what you find in general dictionaries.

A thesaurus lists words that are closely related in meaning. Using one can
help when you want to avoid repeating yourself, or when you find yourself
fumbling for a word that's on the tip of your tongue. But it's important to
remember the difference between denotative and connotative meanings. A
word's denotation is its primary or "dictionary" meaning. Its connotations are
any associations that the word may suggest; connotations may not be as exact
as the denotations, but they are part of the impression the word conveys. If
you examine a list of "synonyms" in a thesaurus, you will see that even words
with similar meanings can have dramatically different connotations. For example, alongside the word *indifferent* your thesaurus may give the following: *neutral*, *unconcerned*, *careless*, *easy-going*, *unambitious*, and *half-hearted*. Imagine
the different impressions you would create if you chose one or the other of
those words to complete this sentence: "Questioned about the experiment's

chance of success, he was _____ in his response." Remember that a reader may react to the suggestive meaning of a word as much as to its "dictionary" meaning—and that suggestive meanings can change over time (see Chapter 13). These comments also apply to the thesauruses included in computer programs. Unless you are absolutely certain that the word suggested is the one you want, it's best to check in a good dictionary and confirm the meaning of the word.

Completed copy

Again, don't forget to number each draft so that you make it easy to differentiate one from the other. Once you have settled on the final version, part of the file name should indicate that fact: e.g., <file name.FIN>.

PROFESSIONAL PRACTICE

Although this book is intended primarily for students, its advice on how to present your ideas also will be useful to you in the work world. Indeed, the ability to convey your thoughts to others in a clear and concise fashion could be the critical edge you need to obtain and hold a job, and to achieve advancement. The contents of this book, then, are *not* just suggestions to help you think, organize, research, and present findings for academic course work: they also are aids to learning skills that will be important later, in whatever field(s) of endeavour you eventually find yourself.

NOTES

1   The authoritative multi-volume dictionary of English words: *Oxford English Dictionary*, 2nd ed. (Oxford: Oxford University Press, 1989), plus the Additions Series volumes (various dates), available online, and, as the 3rd ed., on CD-ROM (2005). *The Concise Oxford English Dictionary*, 11th ed., revised (Oxford University Press, 2008), a single volume, is the most popular dictionary of its kind around the world. It is available also on CD-ROM. Regionally pertinent dictionaries, with localisms, include: *The Canadian Oxford Dictionary*, 2nd ed. (Toronto: Oxford University Press, 2004); *The New Oxford American Dictionary*, 2nd ed. (New York: Oxford University Press, 2005); *Random House Webster's College Dictionary with CD-ROM* (New York: Random House, 2005); *American Heritage Dictionary*, 4th ed. (Boston: Houghton Mifflin, 2000); *The Australian Modern Oxford Dictionary*, 3rd ed. (Melbourne: Oxford University Press, 2007); and *The New Zealand Pocket Oxford Dictionary*, 3rd ed. (Auckland: Oxford University Press, 2005).

2   Pertinent disciplinary and interdisciplinary dictionaries include Susan Mayhew, *A Dictionary of Geography*, 4th ed. (Oxford: Oxford University Press, 2009); Ronald J. Johnston, et al., eds. *The Dictionary of Human Geography*, 4th ed. (Oxford: Blackwell, 2000); David Thomas and Andrew Goudie, eds., *The Dictionary of Physical Geography*, 3rd ed. (Oxford: Blackwell, 2000); John B. Whittow, *The Penguin Dictionary of Physical Geography*, 2nd ed. (London: Penguin, 2000); Audrey N. Clark, *The Penguin Dictionary of Geography*, 3rd ed. (London: Penguin, 2003); Linda McDowell and J.P. Sharp, eds., *A Feminist Dictionary of Human Geography* (London: Arnold, 1999); B. Delijska, comp., *Elsevier's Dictionary of Geographical Information Systems in English, German, French and Russian* (London: Elsevier, 2002); Association for Geographic Information and the University of Edinburgh Department of Geography, *GIS Dictionary* (online: http://www.geo.ed.ac.uk/agidict/welcome.html); Philip Kearey, *The New Penguin Dictionary of Geology* (London: Penguin, 2001); S.P. Parker, ed., *McGraw-Hill Dictionary of Earth Sciences,* 2nd ed. (New York: McGraw-Hill, 2003); Andrew Porteous, *Dictionary of Environmental Science and Technology*, 4th ed. (New York: Wiley, 2008); Michael Allaby, *A Dictionary of Ecology*, 3rd ed. (Oxford: Oxford University Press, 2006); Michael Allaby, *A Dictionary of Plant Sciences*, 2nd ed., rev. (Oxford: Oxford University Press, 2006); Elizabeth Martin and Robert S. Hine, eds., *A Dictionary of Biology*, 6th ed. (Oxford: Oxford University Press, 2008); Chris Park, *A Dictionary of Environment and Conservation* (Oxford: Oxford University Press, 2008); and Michael Allaby, *Dictionary of Earth Sciences*, 3rd ed. (Oxford: Oxford University Press, 2008).

CHAPTER 2

Doing Research

Topics discussed in this chapter include sources of knowledge; induction and deduction; the spectrum from objective detachment to subjective involvement; moral and ethical issues; hypotheses; three types of sources; and the importance of the library, the Internet, and other computer-based resources.

RESEARCH

Whether you are a social or a physical scientist, or a genuine combination of the two—your orientation may have to change from course to course—it's vitally important that you learn how to "do" research. The *Oxford English Reference Dictionary* defines research as "systematic exploration and study in order to establish facts and reach new conclusions." The process of doing research can take several forms. Typically, though, a synthesis of what is already known leads to the formation of a new question, or set of questions, and perhaps to the advancement of one or more hypotheses. After that stage, data is obtained and examined, using pre-determined methods. On the basis of the data examination, a conclusion is drawn with respect to the basic question(s) or hypotheses. The nature of the exploration and the results you achieve depends on your particular purpose, topic, data, and methods. In all cases, however, research consists in the *purposeful* setting out to understand an event, pattern, or process, in which new data, or old data under new circumstances, are examined critically and, perhaps, tested in some manner.

Asking questions

The ability to ask intelligent questions is one of the most important, though often underrated, skills that you can develop for any work, in or out of postsecondary studies. All research is based on questions. Even if you are given explicit instructions by your professor, lab book, or course guide, it's still important to pose questions about what you're doing, how, and why. More independent assignments expect you to ask questions in order to define what you intend to do and how you will go about it.

The scale you select for your study will help you identify the processes to be examined and the kinds of data you will need. For example, a student of fluvial processes will ask different questions about a whole river basin than about a 100-metre stretch of a stream, while a student of political geography will be attentive to different concerns depending on the size of the region in question. A large-scale (small area) study will lead to different questions, types of data analysis, and insights than will a small-scale (large area) one.

To be useful, any question should be *limited, unified,* and *exact* (for tips on developing a hypothesis see pp. 28–9). Students in geography and other environmental programs often approach their work through a five- or six-question formula: *what? where? when? why? how?* and, in some instances, *who?* For a project in historical geography, for example, you might ask any number of questions: "What contrasts within settlement system *X* can be found for the decades 1850, 1860, and 1870? What was the nature of the land before the area was settled? Who were the settlers? Where did they come from? How did the people perceive, utilize, and organize the land? How was the land affected over time? When did patterns of use, and their effects, change, and why? What sources can I use?"

By contrast, as a student in a soils course you might ask: "What is the soil type in place *Y*? How does its profile contrast with that in place *Z*? What processes were involved in the formation of the soil? How, when, and why did these processes effect change? Was human action involved in disrupting certain parts of the soil profiles? If so, who were the people and what was the level and type of technology involved? What are the consequences of my findings for the area if it is to be used for housing (or a horse farm, or a shopping plaza, etc.)?" The nature of your questions depends in large measure on the specific subject of your work. More broadly, however, the questions you derive will depend on the type of knowledge you draw upon.

Sources of knowledge

Three elementary sources of human knowledge are tradition, authority, and personal experience. *Traditional* knowledge is specific to a particular society and is passed from one generation to another. Thus, it is important to remember that the "truth" accepted as given by a particular "tradition" is not universal. A people's cultural traditions, as they are shared and remade through the generations, are central to an understanding of that people's present orientation, and are an important aspect of many types of research, especially in cultural, political, and historical geography. Traditional knowledge is considered to be less important in the physical or biological sciences (except with respect to topics such as indigenous peoples' knowledge and use of plants and wildlife) than in the social sciences.

Like tradition, *authority* usually has a cultural basis. People in positions of authority—"experts" of various kinds (political, religious, economic, educational, medical, etc.)—often are able to "guide" others to what they see as "the truth." Their expertise frequently goes unchallenged; in some cases, it cannot be challenged, because of the threat of reprisals. But such authorities are not infallible, especially if their knowledge is based principally or only on personal experience. It is essential for all academics—students as well as faculty—to examine sources of authority, whether people or institutions, and to be ready to challenge "expert" research, especially if their own work seeks to build on it. In fact, the principle of academic freedom was established in response to the risk taken by researchers who sought understanding and "truth" that put them at odds with "the authorities."

Finally, *personal experience* should not be discounted. All people have the ability to recognize regularities, to generalize, and to make predictions on the basis of personal observations. Experience, especially "trial and error," can lead to knowledge, even if a "discovery" is simply being made afresh for the millionth time; sometimes, if we reflect on such experience, it can lead to fresh insights. As a means of gaining knowledge, however, trial and error tends to be unsystematic, haphazard, and inefficient. For all practical purposes, personal experience may be inaccessible, unless a researcher purposefully and systematically sets out to observe, record, and understand it.

These sources of knowledge—tradition, authority, and personal experience—both reflect and affect people's ways of thinking, perceiving, and doing things, and can themselves become subject matter for students in geography and the environmental sciences. At the same time, it is important for all researchers—even those engaged in apparently objective/quantitative research—to be aware of the way their own traditions, authorities, or personal experience may cloud their view. The old saying "know thyself" applies here; unless researchers are aware of the impact these elemental sources of knowledge can have, ignorance or unthinking bias may adversely affect the way they approach and carry out their research.

The spectrum of approaches

Many researchers who adhere to the Western scientific tradition explicitly seek to avoid relying on tradition, authority, or personal experience as sources of knowledge. Instead, they seek knowledge and understanding by logical reasoning, which can take two contrasting forms: induction and deduction.

Induction proceeds from particular cases through an ordering of information to generalization and, ultimately, universal statements: that is, from the specific to laws, theories, and explanation. The quality of knowledge arrived at through inductive reasoning depends on the representativeness of the specific cases used as the bases for generalization. The reasoning process itself does not include any mechanisms for evaluating the representativeness of the cases considered, and has no built-in checks or controls.

By contrast, *deduction* proceeds from a general *a priori* premise (presented as a hypothesis), through experimentation and verification to explanatory statements about particular events. Deductive reasoning itself is not the source of new information; rather, it is an approach to illuminating relationships as the researcher moves from the general to the specific. Deductive reasoning depends on acceptance of the initial generalizations as the assumed truth. As techniques for solving problems, both inductive and deductive reasoning have limitations. Consequently, most approaches to scientific research involve some combination of the two. One such approach is called the scientific method.

The *scientific method* combines important aspects of deductive and inductive reasoning to create a system for obtaining knowledge. One important feature of the scientific method is that it provides for self-evaluation, using a system of checks and balances to minimize the possibility that the researcher's emotions or biases will affect the conclusions. This systematic approach to inquiry relies on a general set of orderly, disciplined procedures used to acquire dependable and useful information. The researcher progresses logically through a series of steps, according to a pre-established plan:

1. formulating and delimiting the problem
2. thoroughly reviewing the related literature
3. developing a theoretical framework
4. formulating one or more hypotheses
5. selecting a research design
6. specifying the object or population for study
7. developing a plan for data collection
8. conducting a pilot study and making revisions
9. selecting the sample
10. collecting the data
11. preparing the data for analysis
12. analyzing the data
13. interpreting the results
14. sharing the findings with others

The researcher who faithfully follows the scientific method is expected to be a neutral observer who records and analyzes empirical phenomena (for example, a soil's chemical characteristics, a vegetation pattern, the flow of a rock glacier, settlement patterns, traffic flows) according to the tenets of science. Individual phenomena or events are accepted as exemplars of general laws (such as the physical law of gravity, or the gravity model used for measuring the volume of trade between two places, or the laws embedded in central place theory)—laws that are held to be invariable across time and space. Results produced following this approach, emphasizing *a priori* theory, empirically measurable evidence, mathematical modelling, and statistical analysis, are said to be objective. Accordingly, any other researcher should be able to replicate this work (that is, your work, if you are the researcher).

The stress on empiricism in the scientific method means that the person doing the research is removed from the investigation. Recently, however, strong reactions against this emphasis on detachment have arisen in the social sciences, including much of human geography. As different ideas have emerged concerning what is "real," human geographers and others involved with environmental studies have been exploring other theoretical approaches to research, including *humanism*, *phenomenology*, *existentialism*, *semiotics* and *iconography*, *marxism*, *structuration* and *structuralism*, *postmodernism*, *deconstructionism*, *anti-colonialism*, and *feminism*. Only two examples are noted here, but many worthwhile books are available that offer overviews of the various approaches.[1]

If empiricism/positivism lies toward one end of a spectrum, *humanism* lies toward the other end. The humanistic researcher is concerned with the role of human experience and meaning in understanding people's relationships with their geographical environments. The researcher seeks to appreciate, describe, and interpret human awareness and action as they both create and are created by such geographical qualities as space and place, nature and landscape, the built environment, and "geographies of the mind." The humanistic vantage point is generally *not* that of a dispassionate, neutral observer but, as far as possible, that of an insider, one who may be involved in the process itself, who seeks to understand attitudes, values, and perceptions. If you are working from a humanistic perspective, you will be concerned particularly with understanding the three forms of knowledge identified earlier—tradition, authority, and personal experience—and your writing style likely will be a good deal more subjective and personal than it would be for, say, a science lab.

Feminist geographical inquiry—practised by male as well as female researchers—is explicit in its use of qualitative research methods. Here the goal is direct engagement of "real-life" individuals within their "lived reality." Researchers strive to ensure that the research process is beneficial not only to the researcher but to the subject; in fact, the research is expected to contribute to the subject's empowerment. In relating to their subjects, researchers explicitly must address differences in constructions of gender, material circumstances, life experiences, and class, as well as perceived or actual differences in power and knowledge. Writing from a feminist research perspective necessarily will lead you to take a more immediate, or personal, approach than you can when you use the dispassionate scientific method.

HANDLING THE SPECTRUM OF APPROACHES

There are many approaches to research, then, ranging across the spectrum from extreme objectivity (quantitative) at one end to extreme subjectivity (qualitative) at the other, with many variations in between. The goal of objectivity in research—through application of the scientific method—is a powerful guide to critical thought. But a strictly objective/quantified method also can be a negative filter if it means ignoring other, more subjective kinds of research and explanation. Valid arguments can be made for research to be undertaken at both ends of the spectrum. How should you deal with the tension between them? By remaining open to the various approaches. Think about the different questions that each approach may raise—and be aware that no single approach necessarily will lead to *the* answer on everything. Different approaches to research are useful at different times and for different types of research. Students in geography and other environmental sciences need to become familiar with the range of approaches to research, and with the varying ways of integrating the objective with the subjective. With practice you will learn how to approach a wide range of assignments.

Ethical issues: Questions and getting necessary permissions

Every research approach entails ethical questions that often are central to the research agenda. Why should this be the case? Donald Meinig, a prominent geographer, reminds us that "those who have a deep fascination for the earth . . . must have a special concern for the care of the earth."[2] This "special concern" applies to the way we do research, by having sensitivity for any impact our research may have on the natural environment and the people we are to involve in our research. Two questions are essential: "Why is this research being done?" and "What is its end purpose?" Implicit in these questions, and the answers you give, are ethical issues that you must deal with. In some instances, you must get permission to undertake the research, as noted below. Before proceeding, however, a word of caution: remember that when you are doing research that in any way involves others, you are a representative of your institution and any lasting impression you may leave could have positive or negative ramifications for the institution and for other researchers who will conduct research after you.

PHYSICAL AND BIOLOGICAL SCIENCES

Respect the natural environment, including the terrain and all flora and fauna. Ask yourself if there could be any short- or long-term harmful consequences for the research site from the proposed research. How will you seek to minimize these effects? Can you reformulate your approach to avoid any negative impact? Can you identify long-term benefits from your work? How do the benefits weigh against any possible negative consequences? Be sure to identify these in your research proposal. Your department may have an environmental research ethics committee that will scrutinize your research proposal before you are permitted to proceed. You will be asked to discuss the following points:

- the purpose or rationale for doing the study
- the study site (why there, and who must be approached for approval to work there)
- the type of data to be gathered
- the research methodology
- any possible risks (to the environment, and also to yourself)
- possible consequences (good or bad) of your research
- anticipated outcomes

If your research involves animals, you should anticipate requesting approval for your project from a use of animals in research committee, at the department level or, possibly, at the university level. You need to determine if you will be required to get permission from a government, a landowner, or a First Nation to do the research on their land. Before making an approach to get permission, you will need to prepare an outline of your research plan in which you highlight what will happen at the site, how you will analyze the data gained, possible outcomes, and how you will share the research findings, so the person or agency responsible for giving permission is fully aware of your plans.

HUMAN GEOGRAPHY WITHOUT INTERVIEWING PEOPLE

This title implies research on artifacts in the cultural landscape or the use of data obtained by observing people's actions or from examining documents. The provisos identified above apply in these instances too. If you are permitted access to property or documents, take care not to change or damage them. Remain true to what is revealed from your observations and by your analysis.

RESEARCH INVOLVING HUMAN SUBJECTS

Research involving humans inevitably has a subjective dimension, explicit or implicit. For example, suppose that you intend to develop, apply, and analyze a questionnaire using the scientific method. In such a case, you will attempt to "remove" the subjective aspect by making the questions as value-free as possible, using strict sampling methods, and employing people not involved with the research design and the testing of data to approach the people who are to respond to the questions. An "objectively" structured questionnaire may not be appropriate in all cases, however. If you want to find out about people's attitudes, values, and perceptions, an explicitly "subjective" questionnaire, or perhaps open-ended interviewing, with the data being examined using qualitative techniques, may be more appropriate.

By posing a question, you place that question into the mind of someone who may never have thought of it before. Thus, you have to think carefully about the purpose and appropriateness of your questions. Furthermore, no matter what approach you take, it is imperative that you realize that the questions you ask inevitably reflect your own attitudes and values. This means you must be very careful when choosing what it is you truly want to ask because you don't want to unthinkingly bias the respondent. Researchers whose research is fundamentally subjective in nature must be intensely aware of the way their own perspectives may affect their work, and they often invite participants themselves to verify the findings—a step that strict followers of the scientific method may be reluctant to accept.

All upper-level research undertaken by honours students, graduate students, and faculty that involves human subjects is now vetted by research ethics committees (which, depending on your research level, may be at the departmental, college or university level). This requirement always applies, even if the research is funded by external research agencies (some of which have their own research ethics committees and stated requirements). Some lower-level research, whether by single students or by several, in a class project, also may have such a requirement, vetted by a departmental or university committee. The lower-level requirement may not apply everywhere, so check your institution's policies. A useful source to consult about the ethical conduct of research is the joint policy statement by the three research funding councils in Canada: the *Tri-Council Policy Statement: Ethical Conduct for Research Involving Humans* (1998, with 2000, 2002, and 2005 amendments). This document is available electronically at (http://www.pre.ethics.gc.ca/english/policystatement/policystatement.cfm).

Some basic principles apply when doing research involving others:

- Get permission to do any research at a site that is owned or controlled by others, by first being totally open about what you propose to do and why.
- If you are to involve people in your research, be open and honest with them about the purpose and nature of your research so they understand why they are being asked to participate. When appropriate, or as required by your institution's ethics committee(s), give them access to a written description of your research purpose and design.
- Obtain freely-given informed consent prior to involving people in your research. This may be done in writing (on a formal consent form, or at the top of the interview questionnaire) or obtained verbally (using a tape recorder, or recorded in your field notes—in which the person's initials may be recorded against your notation that permission to proceed has been granted). In all cases, be sure you also record the date.
- Ensure that each person understands that he or she has the right to withdraw from your study at any time, without explanation or recrimination. If a person does withdraw, identify in advance how you will deal with the information given to you up to that point. If you intend to retain and use the information provided, you need to indicate your intention at the time you seek informed consent from your participants.
- Ask your respondents if they wish to remain anonymous. Know in advance, within your research design, how you will deal with the information gained from a person who wishes to remain anonymous. If a person wants to remain anonymous, identify how you plan to identify his or her input, and ask if this is okay.
- Inform all of your participants how you plan to use the material gathered (e.g., for an undergraduate essay, or a graduate thesis) and who will have access to it (e.g., your supervisor, an examining com-

mittee, or, if it is a thesis, anyone who may see the work in the library).

- Treat your research subjects—and the research site—with respect.
- Be sensitive to gender and cultural differences.
- Always be courteous, even when opinions quite contrary to your own are expressed.

If you are required to complete a form for review by a research ethics committee, be sure to build in time for the review to take place. Reviews may require many weeks to complete, including time for the committee to review your application, as well as time for you to respond to any questions the committee may ask, and subsequent follow-up queries from the committee. If in doubt, speak with your faculty adviser.

Stating your purpose and developing a hypothesis

As Chapter 1 pointed out, you need to develop a clear statement of purpose for all the research and writing you do. In qualitative research this statement often is framed as a question, but in quantitative research it normally takes the form of a hypothesis. A hypothesis is a working thesis that translates your problem statement or question into a prediction of expected outcomes. By identifying the line of argument that you intend to follow, a hypothesis helps to make your research selective. In research structured by the scientific method, you must hold your hypothesis constant throughout the research process.

Whether you start with a question or a hypothesis, you must return to it at the end of your paper: if you started with a question, you must answer it; if you started with a hypothesis, you must state whether your findings did or did not support it. Whatever form your statement of purpose takes, it should be *limited*, *unified*, and *exact*. The following advice refers specifically to hypotheses, but it also applies to the formation of questions.

MAKE IT LIMITED

A limited hypothesis is one that is narrow enough to be workable. Suppose, for example, that your general subject is land-use planning in Canada: you must limit the subject in some way and create a line of argument for which you can supply adequate supporting evidence. Following the analytic questioning process, you might find that you want to restrict it by time. Imagine that you start with a question: "What role did government play in the development of land-use planning from 1950 to 1960?" This question can be changed into a statement: "Government action led to the development of land-use planning during the decade 1950 to 1960." Or you might prefer to limit it more precisely, by theme and location: "Legislation introduced by the government of British Columbia led to the development of land-use planning during the decade 1950–1960." In short, make sure your topic and purpose statement are restricted enough for you to explore the issues in the depth required in order to draw supportable conclusions.

MAKE IT UNIFIED

To be unified, your paper must have one controlling idea—a hypothesis or key question with which it will both begin and end. Beware of the double-headed hypothesis: "In its term of office the government introduced many social programs, but its downfall was due to its nuclear energy policy." What is the controlling idea here? Social programs, or the reason for the downfall of the government? The essay should focus on one or the other. It is possible to explore two or more related ideas in a paper, but only if one of them is clearly in control, with all other ideas subordinated to it.

MAKE IT EXACT

It is important to avoid vague terms such as *interesting* and *significant,* as in "Cemeteries are interesting places as sources of data for reconstructing settlement history." Does *interesting* mean *curious* or *good* or *exciting* or *likeable* or *important* or *worthwhile,* or . . . ? Cemeteries are *useful* places as sources of data? Yes, they are. They may, *in addition,* be interesting—but it is their usefulness that is important here. Use words that state precisely what you intend to do.

SOURCES

Research is a building process: researchers always build on what they and others have done. As the previous chapter noted, there are three kinds of sources: primary, secondary, and tertiary. Primary sources are the sorts of data that you gather for yourself, whether in the field (see Chapter 9) or in the lab (Chapter 6), though for some types of research your primary material may have been produced by others and published in books, journals, reports, newspapers, letters, or elsewhere. For all of your work, you will need to consult secondary and tertiary sources. The remainder of this chapter provides some suggestions for finding and using sources in libraries and on the Internet.

Libraries

You must become familiar with your institution's libraries. (The plural reflects the fact that in many institutions the map library and some subject libraries are separate from the main library.)

Don't put off going to the main library until near the end of the term; in fact, it's a good idea to renew your acquaintance with it at the start of every academic year. And since you likely will be taking courses in different disciplines each term or semester, at the start of each, check to see where the key areas in the library are for the courses in which you are registered. The importance of getting to know your way around the library can't be stressed enough. You don't want to be so overwhelmed by its size and complexity that you either scrimp on required research or waste time and energy trying to find information.

Disciplinary "fact sheets," which highlight key sources and their locations, may be available from the library. If you need help, ask for it. Librarians will be glad to show you the bibliographies and guides to the literature, dictionaries, encyclopedias, directories, atlases, gazetteers, handbooks, databases, CDs, DVDs, videos, and other materials for your field of study. Some of these sources will be available online; others will be found only on the library shelves.

BROWSING

A useful way to spend some free moments is to wander around among the library's shelves and see what is there. Learn the cataloguing codes for the subject areas—and shelves—covered in your courses and browse accordingly. As you wander, titles and authors you have heard mentioned in lectures or read about in texts are likely to jump out at you. "Citation tracing"—looking for items identified in your readings—can be rewarding. A quick glance through the table of contents of a book or a journal will indicate whether there is something in it you should read or that you should at least note for later examination. Don't ignore the special collections located in various parts of the library or in an associated centre.

SPECIAL COLLECTIONS

The *Reference section* is a "must know" area. Often it is located in an easily accessed area on the main floor of the library. Dictionaries, encyclopedias, bibliographies, various versions of *Who's Who* and other biographical sources, and many other publications you can check readily for specific information are located here. A wide collection of reference materials also is available to you via the library's website, but they may not include what is located on the library shelves.

Newspapers are located in a designated area in the library. Some are in hard copy, delivered daily. They usually are arranged in alphabetical order by title. Older material from these and numerous other newspapers is available on microfiche. A list of titles held by the library is identified in the catalogue and, possibly, in a special link page in the library's website. The library's website likely also includes "current newspaper databases" and "historical newspaper databases," with links you can use to see electronic copies of local, national, and international newspapers. If your work involves historical reconstructions, you may have to go to provincial, state, or national archives to see certain newspapers.

Sound recordings, video recordings, and motion pictures are available to you in another special collections area in the library. You may be able to view or hear some of this material in the library, or you may be able to borrow it, likely on a short-term (such as a two-hour, or one day) basis. Usually you can borrow items for use in a class presentation, but you must plan ahead to arrange a special loan.

The *Government Documents* section of the library is important for some subjects. Located here are materials—including some "fugitive" (hard to find) documents—from national and more local level governments and quasi-governmental agencies, from many countries. Also located here are documents from such non-governmental agencies as the United Nations and its many parts (including, for example, the Food and Agriculture Organization [FAO], the World Health Organization [WHO], the United Nations Development Programme [UNDP], the United Nations Environment Programme [UNEP], and the Economic Commission for Latin America and the Caribbean [ECLAC]), the World Bank, and the Organisation for Economic Co-operation and Development (OECD). Materials usually are arranged alphabetically by country or organization. Only some of the data available here may be available online.

Rare Book and Archival Collections in your institution's library may include old (e.g., early nineteenth century) books, unique papers (including diaries and letters) by particular individuals or concerning famous people, or collections of documents from specific organizations. Now there are many archival and rare book library sites available on the Internet, including:

- Archives Canada: http://www.archivescanada.ca
- U.S. Library of Congress: http://www.loc.gov/rr/rarebook
- British Library: http://www.bl.uk/collections/treasures/digitisation.html

DATABASES AND GEOGRAPHIC INFORMATION SYSTEMS

Most colleges and universities now provide amalgamated statistical data resources in a single centre that may be associated with the main library. Such a centre gives you access to myriad data sets and statistical consultations. Some of the data may have been generated by your institution's faculty (and perhaps by senior undergraduates and graduate students), and will be obtainable from just the one source. Other sources are available online, including an increasing variety of governmental and non-governmental statistical data on environmental issues, population, international trade, and much more. Thousands of databases now exist, covering every conceivable topic, including some that are readily adaptable to Geographic Information Systems (GIS). GIS services often are provided by staff in a university data resource centre, which may act as a central repository for GIS data available on campus. Pertinent data also may be located in the Government Documents section or the Maps Division of the library, and in a library or special collection in the Department of Geography.

You may be able to create your own personal database (using, for example, RefWorks) by importing references from text files or online databases, which you then can use when writing papers, including automatically formatting and creating the bibliography.

BOOKS

Books are found in two areas: on the new-book shelf, which is changed frequently (perhaps weekly), and on the general shelves. Depending on the courses you are taking, you will need to become familiar with a variety of source materials in the biological, geographical, physical, and social sciences, and, for some, the humanities as well. Do *not* expect to find all you need online. Accordingly, at the start of each term, find out where the books on the specific subjects you are taking are located: for instance, an ecology section under biology; historical geography under history and geography; resources and resource management under geography, economics, and agriculture; urban issues under architecture, geography, economics, sociology, and history; soils under agriculture, geography, and engineering; geomorphology under geography and geology; climatology and meteorology under those headings as well as under physics, geography, and agriculture; landscape interpretation under architecture, geography, literature, and art; environmental perception under geography, psychology, and urban planning; and so on. Don't forget to examine the many thematic and regional geography sections in the library, which always are usefully varied in their contents.

Since many books have not been refereed, you should be careful when you refer to them as possible sources. Generally, a refereed book is one published by a university press, which means several external readers critically evaluated the book before the press decided to publish it. This does not mean that it is "right" but it does mean that it has passed a test for quality and accuracy. Some commercial presses also publish refereed academic books, but it is important to consider the source of any book you might choose to cite.

The library system provides links to electronic book (e-book) collections covering a wide range of subject areas. Some e-books are available only through the library, due to special agreements. Check to see if the library's website has a special page which identifies the material—and provides links for you to use.

As suggested above, you can save yourself precious time and pressure during mid- and late-term by visiting the library early in each term, well before assignment deadlines. Don't wait until the last minute to get important library sources: otherwise—horrible thought—you may find that the key work you need has been signed out by someone else or that it is available to you only through an interlibrary loan. Plan ahead, and give yourself enough time not only to use the books, but for a "hold" notice to take effect in the event that you must wait.

JOURNALS

Journals normally are located in two sections: recent issues (sometimes containing a year's run) and back-runs (often covering all the years since a journal began publication). You are expected to become acquainted with literature that appears in learned journals.[3]

What makes these journals different from popular publications? Researchers seek to have their findings published in major refereed journals. This means that a paper is published only if it is deemed to be of value. This is determined by the journal's editor, or editorial team, seeking the reactions of two or three leading researchers in a field of study, who anonymously critique a submitted paper and provide a recommendation to publish, reconsider after revisions, or reject. The author's name and the names of the other readers are kept from each reader so as to reduce the chance of personal biases intruding into the evaluation process. Furthermore, the readers' names are kept from the author; thus, in the acknowledgements an author may include with the article, you will see "anonymous reviewers" being thanked for their input. Finally, look for the "submitted" and "revised" and "acceptance" dates that may be identified.

It is daunting to enter the journal sections of libraries—there are so many publications to choose from. You will want to select the journals most pertinent to your courses. However, be aware that there are five major categories of journals: *general, specialist thematic, regional professional, pedagogical,* and *popular* (see Appendix II).[4]

Your library has many journals on the shelves or available online via library links. Get to know what is available to you, so you know the limits of the local library's resources. Be aware that you can get some journal articles from libraries at other institutions. Planning ahead will be essential if, through your library, you need to obtain a hard copy of a journal article from another institution's collection; however, since some article delivery systems are restricted to faculty, staff, and graduate students, you may need to consult your instructor or TA for help in this regard.

How do you locate journal articles? First, be aware that it takes time and effort to find out what is available in journals. Even if you know the title of an article you want (perhaps from a listing in your instructor's reading list or from a bibliographic entry in your textbook), you won't find the article listed in the library's catalogue—although you can use the catalogue to determine whether or not the library carries the journal, and, if so, discover the call number. Often, though, you won't have a specific article title or author to search for. There are several ways to identify journals that are potential sources for pertinent material on specific subjects. Your instructor may guide you to certain journals, as may bibliographies in your text or other pertinent publications (including those that appear in notes 1, 5, and 6 at the end of this chapter). If a particular journal is in the library, you can browse through the table of contents of several issues and read the abstracts that may be published with the articles. Such abstracts may appear at the top of the article, under the title, or (perhaps on a back page) in a group of abstracts for all of the articles published in the same issue. Alternatively, you can glance through a journal's annual index, usually published in the last issue of each volume year; note that this may not coincide with the calendar year. However, browsing in this manner is an inefficient way to search for articles on specific topics, especially if you are pressed for time. Another approach is to explore indexes and abstracts that are published in separate volumes or are available online.

Your institution's library may subscribe to online journal services that some commercial publishers host on their ".com" sites. These publishers make selected high quality, peer reviewed journal articles available in PDF format that your library's subscription enables you to download from their websites.

For example, in mid-2008, publishers Wiley InterScience and Blackwell Synergy merged their online journal systems to provide registered users with access to 1,400 journals, about 3 million articles, and other reference and data sources (www.interscience.wiley.com). Online tutorials are available to help you understand how to search this site; other publishers provide similar sites and assistance.

INDEXES AND ABSTRACTS

Indexes and abstracts provide an organized way to find books and, of special use, research articles published in journals. Some list chapters within edited book collections. Librarians can help you identify the aids most appropriate to your studies, available to you either in the library collection or on the Net. Some of the most useful and pertinent journals are included in Appendix II. You will have to use the library's account to obtain bibliographic, abstract, and full-text information from many online databases. This is because some databases, and most academic journals available online, are accessible to you only through a licensing agreement between the publisher and your institution.

Your library also may have on its shelves *Geographical Abstracts: Physical Geography*, *Geographical Abstracts: Human Geography*, *Geological Abstracts*, *Ecological Abstracts*, and *International Development Abstracts*. Other compilations include, in hard copy and sometimes also—or only—online, *Biological Abstracts*, *Social Science Abstracts*, *Sociological Abstracts*, *Humanities Abstracts*, *Oceanographic Literature Review*, *EconLit*, *Readers' Guide Abstracts* and *Readers' Guide to Periodical Literature*, *Psychological Abstracts* and *PsycINFO*, *MEDLINE*, *TOXNET*, *Web of Science*, *Century of Science*, and the American Geographical Society Library's *Current Geographical Publications* and *GeoBib Search*. For upper-level studies, you may have occasion to refer to *Dissertation Abstracts* (available online).

CHECKING ON AUTHORS

If you want to place a particular work into the broader context of its author's various writings and accomplishments, there are many sources you can use to find biographical information. You can check *Books In Print* for particular authors and your library's catalogue—or, online, the library catalogues at such major research institutions as the University of Chicago and the University of Cambridge. Be sure to use the advanced (Boolean) search function. Other sources include the many (national, international, and subject-specific) versions of *Who's Who*, the *Dictionary of International Biography*

(Cambridge: International Biographical Centre, many editions), specialized dictionaries and encyclopedias,[5] as well as biographies and autobiographies.[6]

CATALOGUING SYSTEMS

How, specifically, do you locate the sources that are in the library? You will see a "call number" associated with an entry in the catalogue. The principal cataloguing systems are the *Library of Congress* system (used in most university libraries) and the *Dewey Decimal* system (used by many public libraries). Find out which system your institution's library uses. Both systems include many sub-disciplines within each category. When searching for books and journals on the shelves, you will save time if you learn which of the major categories applies most directly to your research needs.

USING THE CATALOGUE

You can't just stumble into a library and "start." You have to proceed with a plan. The "information sphere" is large, complex, and often confusing, so you need to take time to think calmly and rationally about the information you need. Start by thinking about your purpose and what you will need to do to complete the job. It can be helpful to write down the various research steps required. Pose some questions for yourself—in addition to those posed on p. 3, it also is important to think about how many authors you should consider.

Before you can focus on tackling your central question or hypothesis, you need to get a sense of the context required to frame it. An important part of the "framing" process is identifying the data you will use. Will they be original data that you have gathered in the field, or information located in books, journals, census reports, magazines, newspapers, published or online data sets, CD-ROMs, or any of many online sources? How will you find the material you need? You might start by checking the bibliographies in your textbook, reading the lists provided by instructors, and scanning footnotes and endnotes in key books and journal articles. You also will want to use the library catalogue. However, if you choose to search by book title, you may discover that a title is not always a reliable guide to a book's content. To find useful source material, whether in the library's catalogue or online, you will need to make up a list of keywords.

Keywords must be chosen with care, since vague terminology will produce large quantities of non-relevant material. For example, using the term "territory" in a computer-based search will turn up items concerning a particular area (such as the Northwest Territories, or the Northern Territory),

Library of Congress System

A   General Works

B   Philosophy, Psychology, Religion

C   Genealogy

D   History—Europe, Asia, Africa, Oceania

E   History—U.S. General

F   History—U.S. Local, Latin America, Canada

G   Geography, Anthropology, Recreation, Atlases

H   Statistics, Economics, Sociology

J   Political Science

K   Law

L   Education

M   Music

N   Fine Art

P   Languages and Literature

Q   Science

R   Medicine

S   Agriculture

T   Engineering

U   Military Sciences

V   Naval Sciences

Z   Library Science

Dewey Decimal System

011–099   Generalities (including Computing)

100   Philosophy, Psychology

200   Religion

300   Social Sciences

400   Languages

500   Science

600   Technology

700   The Arts

800   Literature

900   Geography, History, Travel

A–Z   (by author) Fiction

international law, personal territory, tourism, terrorism, burials, and thousands of other things, including advertisements! Or, as another example, "self-determination" will turn up items on everything from empowering persons with severe physical impairments to make decisions for themselves, to the efforts of national minorities to gain political independence from the existing state. In other words, it is important to think of additional keywords so you can be precise. Rather than a "basic search," try the "advanced search" mode. For example, you might use "territory" and "geography" or, if too many items appear, you might refine this as "territory" and "political geography." Some catalogues permit you to add an additional word. However, some library catalogues will not permit you to do such a search at all, even in the advanced search mode. You may be limited to one word for the subject, though you may be able to use your subject keyword along with an author's name.

Once you have a list of possible sources, you can rank them from the earliest work to the most recent, or from the most pertinent to the least. Once you have decided which of the sources you wish to use, the next step will be to find them; then you can begin reading and taking notes (see Chapter 3).

Although cataloguing systems vary, as noted above, there are three ways to use them to locate sources:

- By using *known specific information*: *author* (with surname first, followed by the given name or the initials), *title* (of a book or a journal), or (if you don't already know the title of the work or its author) *call number*. Such searches can be quick, but you must know precisely what you are seeking.
- By *subject*. Subject searches are especially useful when you know neither the authors nor the titles you want. Subject searches use *keywords* that you select to search for pertinent book titles. If your keyword is highly specific, you may be able to use just it. But since your chosen word will probably not turn up in all the titles that can be helpful, it is important to try variations, using other keywords.
- By conducting an *advanced search*. Most computerized library catalogues have an advanced search mode, which may permit you to enter several bits of information. Some are limited to title, and/or author, and/or subject, though you may also be able to identify a word you don't want included in the search. These systems can be frustrating because you may not "hit" on what you need. Better systems permit you to use several words, each on a different line; this leads to the greater identification of potential sources. If your uni-

versity library doesn't have this type of advanced search, go online to another university's catalogue and see if what you need is there. Once you have identified something that way you can return to your own institution's catalogue to check if the work is in the collection.

Typically, subject searches are not helpful for finding materials in the Government Documents, Special Collections, or Archives sections of the library. To use these sources, ask the "help desk" librarian or archivist for assistance. If in doubt about how to use your keywords when searching the catalogue, or how to do an advanced search, check to see if there is an online "Library education" or "Help" page, or seek help from a librarian.

Maps as sources

Your institution's map library undoubtedly has myriad loose-leaf maps, map folios, atlases, air photographs, and satellite images, as well as books on cartography, GIS, and related topics. Help may be required to find some of the material, because certain maps, photographs, and satellite images may not be included in the general catalogue. You may have to refer to a particular catalogue, computerized or not. Some material may be grouped and filed under a general heading, in a map cabinet drawer. You will make use of maps that have been compiled and published from various sources. The following criteria may be of use when you are considering maps as sources.[7]

- **Scale.** The scale of the map determines the size of area and amount of detail that can be shown (see Figure 2.1). There are several categories of map scales:
  - *1:1,000–1:2,500* (large scales, small areas; details—lot boundaries, building outlines, road widths, etc., are shown correctly to scale)
  - *1:10,000–1:20,000* (fields, lot lines, and buildings may be shown by conventional symbols, and streets will be named)
  - *1:25,000–1:50,000* (topographical scales, used for surveys of relief —usually shown by contours and spot heights but sometimes by shaded relief)
  - *1:250,000–1:500,000* (progressively more generalized maps of topographic type, with less and less detail)
  - *1:1,000,000 and smaller* (smaller scales, larger areas)
- **Format and media.** Size and layout? Sheet maps? On fiche or computer disks? Blueprints? Hand-drawn? Computer-generated?

Electronic? The answers to these questions will determine whether and how you can use the maps.

- **Currency of the data.** Does the date of the data, and the map, relate to the period you are studying? Be wary of "old" data released in map form years after a survey was done—look for both the date of the data survey and the publication date of the map.
- **Reference structure.** Examine the map for evidence of accuracy. On what reference structure is the map based? Are sources cited? Is the projection named? Is the north arrow shown, and the magnetic variation? Is the scale reference identified?

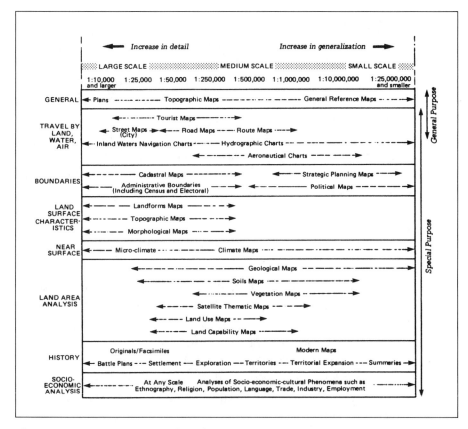

Figure 2.1   Map types and scale ranges

SOURCE: Reproduced with permission of the compiler, B. Farrell, from B. Farrell and A. Desbarats, *Guide for a Small Map Collection,* 2nd ed. (Ottawa: Association of Canadian Map Librarians, 1984), p. 30.

- **Cartographic symbolism.** Do the symbols selected for the categories into which data are sorted follow a graded pattern (e.g., from lesser to greater intensity), or are they confusing? If, say, settlements are identified by size, do the point data clearly reflect their relative sizes (small, medium, large)? If the symbols are non-standard, are they explained in a legend? Do geographical names follow standard authorities?

- **Graphic language.** Does the map successfully convey the data, at an appropriate scale, using clear words and symbols? If the mapped data are complex, involving several sets of information (e.g., rivers, road and field boundary lines, settlements by size and type, settlement names, crop patterns, and political boundaries, or, as another example, data on three variables, such as hydrological balance, precipitation, and temperature), how well does the map deal with the overlapping data? Can you read the locational information and the contrasting areal patterns with ease? If you want to use just some of the information, how easy is it to extract exactly what you need? Would a different map, or maps, be better?

- **Identification.** Is the map titled? Is it part of a series? Is the compiler identified? Is the source of the data cited? Who and where is the publisher and when was the map published?

THE INTERNET

The Internet is an expansive, remarkably diverse, and, at times, frustrating resource. It must be used with care if it is to be of value as a research tool. Hundreds of books exist on how best to use the Internet. What follows is a list of summary pointers: a skeleton guide to the Internet, how to do searches, validating your sources, and selected Internet addresses. (For guidelines on citing Internet and other computer-based information see Chapter 11).

"Google" and "Yahoo" are examples of search engines (robot-like "crawlers") that enable you to find information on the Internet. "Google Scholar" searches specifically for scholarly literature, mostly articles, but including some citations to books. Through http://scholar.google.com you may obtain the full-text of online articles if they are made available freely or provided with permission from academic publishers and professional societies. If an article is protected by copyright, you may find an abstract is available.

How to search online

No one search engine provides complete coverage; speed, database size, relevancy rankings, search results, features, and commands vary from engine to engine, so you probably will want to use several engines in the course of your search. Even if the same keywords are used as you go from one search engine to another, different results will appear. You should be aware that some Internet service providers actually limit the range of sites you may visit. For example, some universities block access to sites from which students might download copyrighted materials, and other providers may block on the basis of content. Most search engines have "help" tutorials you can turn to if necessary. The following points cover some common essentials:

- **Use your keywords.** Start your search with the *keywords* you identified when developing your research plan. Use academic and professional terms as much as possible (e.g., architecture vs. house).
- **Use lower-case letters when typing keywords.** Upper-case letters may restrict your search; if you use lower-case, the search engine will consider sites that use both upper- and lower-case letters.
- **Use a simple search if you are sure the search words are restrictive.** For instance, if you know that you want the "Otago Daily Times," that is all you need to enter, though a search engine may ask you if this is a newspaper (which it is). But be warned: a simple search using a wide-open term, such as "geography" or "biology," is likely to give you millions of sites, from which you will have to choose.
- **Use an advanced search to gain access to the most pertinent sites.** Most search engines permit you to do "advanced" searches. They make it easy for you to do this type of search by having several lines in which you can put what you want, and don't want. You may be permitted to target publications in a particular language, within varying time periods, that have certain usage rights, and with certain types of occurrences.
  1. If the just-mentioned convenient set-up is not present, you may have to use *Boolean* operators and syntax, i.e., AND, OR, and NOT (typed in either upper- or lower-case letters) to link your keywords. In some systems, the minus (–) sign can be used as a substitute for NOT and a plus (+) sign as a substitute for AND.
  2. If you use Boolean operators and syntax, use parentheses to group words: for example, you might search for "caves AND (stalactites NOT stalagmites) AND Tennessee."

3. If you suspect that your keywords are not finding as wide an array of sources as you had hoped, you can carry out a *controlled language search*. A controlled language search uses *pre-existing subject headings*, such as those listed in the five-volume *Library of Congress Subject Headings*.[8]

- **Try alternate spelling.** To expand a search, try variant spellings of certain words: e.g., harbour and harbor; labour and labor; encyclopedia and encyclopaedia.

- **Check to see if there is a site map,** which is like a table of contents. It may quickly give you an idea of how useful the site will be for your needs.

- **If you find a source that is close to what you need, try the "links" or "other sites like this" section as well.** Links to other sites can be helpful. Your library's website may have pertinent links for you to try, too. Similarly, disciplinary websites, set up by specific units (such as your own department), often identify links to other sites of potential interest, and they may well be of special use to you.

- **Don't get stuck on fascinating sites that are, at best, tangential to what you need.** Anyone who has surfed the Web knows how seductive it can be. But you can't afford to spend hours surfing when you really need to focus on research for a specific assignment. Be warned!

- **Save particularly useful websites.** When you find a website you plan to visit again, you can store its URL (Uniform Resource Locator) by clicking on "Favorites" (in Internet Explorer) or "Bookmarks" (in Netscape). By using the stored URL you can quickly return to this site at some other time without having either to remember what might be a very long site URL or to conduct a new search to find the site. Both Netscape and Internet Explorer let you create folders into which you can save the site information. Name the folders so they make sense for your research needs.

  Since all content on the Internet may be moved, or restructured, or deleted, some URLs may not work when you return to them. To help resolve this problem, many scholarly publishers have begun to assign Digital Object Identifiers (DOIs) to journal articles and other documents. Assigned by a registration agency, a DOI is a unique alphanumeric string that provides a persistent link to the Internet location of each item. When a DOI is available (find it displayed prominently on the first page of an article), include it instead of the

URL in your reference (use the copy and paste function to copy the identifier because the DOI string can be long).

- **Record the full URL and the date the site was accessed.** You will need this information for citations within the text of your work and for your list of references.
- **Selected directories may be helpful.** Each directory is selective, so you will need to check several. See, for example, *Librarians' Internet Index* at http://lii.org/, a publicly-funded website updated regularly by librarians, who search for and select the best sites on the basis of the sort of criteria described below under "validating your sources." See also *Virtual Library* at http://vlib.org, for which volunteers from various specialties—but the list is far from being complete—compile lists of key links for particular areas in which they are expert, and each maintainer is responsible for the content of his or her own pages.
- **Know when to give up.** Keep in mind the "law of diminishing returns" and try to recognize when enough is enough.
- **Know when not to use the Internet.** Remember that not everything is available online. A great deal of information can be found only in printed sources. When in doubt, seek guidance from your instructor and reference librarians.

Validating your sources

Your biggest challenge when using the Internet is not always finding what you need. Even if you find many sites that seem to be pertinent, you will want to check if the sites have merit and to decide how trustworthy they might be. Unlike the library, where most of the material available has been critically reviewed, edited, and published by generally reputable outlets, the Internet is unregulated and there are no quality controls. Anyone can publish material online, and many pages are not updated. You will find some gems, but you will also find junk. How can you differentiate? Just as you must use your critical skills in assessing books and journals, so you must assess the value of everything you find on the Net. Asking yourself the following questions will help you establish the integrity, validity, timeliness, and authorship of what you find (see also Table 2.1).

- **Whose material is it?** Who is the author? Who created it? Is the site published by an academic institution? A government or private research centre? A government department? A business? A news agency? An international organization? A group of people with no professional affiliation? An individual? Even if a website is "signed," qualifications may not be provided. Ownership information generally is identified at the top or the bottom of a site (read the URL carefully), or there may be a *home* page in which the information is presented. If there is no such information, be suspicious.

- **Why are the URL and the domain important?** You can use the URL and domain information to check on who wrote the page, how current it is, and the author's credentials. First, an example of a *URL* is: http://www.example.org. Part of this URL is the *domain*: example.org. Now, one way to check the status of a page is to "truncate the URL" by clicking on the URL in the location box, deleting everything after the last part of the *domain* (e.g., delete the text after ".com/" in http://www.whatever.com/notes/file3/other.html), and hitting *enter*. If the page that loads is a reputable source, there is a good chance the document you found is also reputable.

  However, even reputable sources can make errors, so there is no guarantee that the information you discover is accurate. This is one reason why careful reading and assessment of each document is required. Also be careful of vanity publishing, even when a reputable URL is used, for example, when students host their own writing on websites hosted by university domains (see below).

- **Does the domain provide clues about the source?** The domain (top level domain) is a hierarchical framework used to indicate the location of a web page on the network. Common domains in North America have included: .edu (education), .gov (government agency), .org (nonprofit and research organizations), .com (commercial), and .net (network related). In June 2008, however, the Internet Corporation for Assigned Names and Numbers (ICANN) eased existing restrictions to allow a much wider range of domain names. Personal names, trademarked brands for major corporations, city names, and many other domain names may be in existence by as early as 2009, depending on the implementation plan and requirements. ICANN also considered that the largest growth areas for future Internet users will be Asia and Africa, and opened for public comment a proposal to write domain names in non-Latin characters (such as Arabic or Chinese scripts).

The domain name refers to the initial part of a URL (down to the first /) and gives you who "published" a page by putting it on the Web and making it public. Some examples of domains include:

— *country:* e.g., *au* (Australia), *bw* (Botswana), *ca* (Canada), *cn* (China), *de* (Germany), *fr* (France), *in* (India), *jp* (Japan), *ke* (Kenya), *mx* (Mexico), *nl* (Netherlands), *nz* (New Zealand), *ru* (Russia), *sa* (Saudi Arabia), *uk* (U.K.), and *za* (South Africa). Even prior to ICANN's 2008 decision, many country codes were no longer tightly controlled and were subject to misuse, so once again it is important to be thorough in your assessment of each site.

— *type:* education, e.g., *edu* (U.S.), *ac* (U.K.); government, e.g., *gov* (U.S., U.K., and elsewhere), *gc* (Canada), *go* (Kenya); and organizations, e.g., *org, com,* and *net.*

The URLs of personal pages very frequently contain an individual's name (such as "pbrowne") following a tilde (~) or percent sign (%) or the word "users" or "members" or "people." No publisher or domain owner vouches for the information in personal pages, which often contain highly biased personal perspectives. While personal pages are not necessarily "bad," you need to investigate the author carefully. Thinking about why a page may have been put on the web will help you avoid being fooled (and looking foolish in return) or becoming a victim of fraud. You need to consider what sort of intentions the author might have for putting information online—to inform and give facts or data, to explain or persuade, to sell, or to disclose? What "tone" does the author use—is it, for example, humourous, exaggerated, or condescending? Are the arguments distorted or overblown?

• **Is the site vanity or self-publishing?** A vanity website means that the material listed has not been through the peer review process that is intrinsic to scholarship or that it hasn't been disseminated by a trade publisher. Some vanity and short-run, generally privately funded, published works (such as a person's memoirs) may have merit, but this can be determined only by critical evaluation. Many people now put such material on their personal websites. The material may be important only to the author! Even when a site's content seems to be scholarly, as with a person's working papers or information that is derived from a thesis or dissertation, the information may be incomplete—particularly if the site is personal rather than academically affiliated. And, again, some universities permit students to have websites using the institution's address as the first component of the URL; don't be fooled by the domain information into thinking that the source must be sound.

- **Who sponsored the site?** Is the site sponsored, officially or unofficially, by a particular organization (such as an environmental advocacy group), an institution (such as a university or a research unit), a corporation, a government, or a quasi-governmental agency? Is there any evidence of third-party financial or other support or sponsorship? With respect to each of these questions, if they hold true, how might the site's content have been influenced?

- **Is advertising included?** If so, can you determine what impact it may have had on the site's content?

- **Who is the audience?** Try to determine what the purpose of the site is. To whom is the content aimed? Serious researchers? School students? Potential customers? People who enjoy reading the vanity rambling of the particular author?

- **When was the site created, revised, or last updated?** The dates of creation, revision, and last updating generally are given on academically useful sites. Be wary of a site that seems to be great but which has no dates. Dates alone will not mean the site is credible, but it is important to know them. Be skeptical if such information is missing.

- **Is the site credible?** Has the site been reviewed or been given awards? When a site is recognized and given an award by others, this information will be presented. Who gave the award or otherwise reviewed the site? A scientific group? A private agency? Friend "Joe"?

- **What is the quality of the site and of the information it provides?** To answer this question you must assume the critical stance of an academic.
  - Is the site well designed and organized?
  - Is it easy to use?
  - Do the icons clearly represent what is intended and can you move around the site with ease?
  - Is the text well written?
  - Is the content convincing?

- Is it thorough in its coverage of the topic, or is it selective and incomplete?
- How in-depth is the material provided? Is it shallow or is there substance?
- Is the material on the page unique? In what way? How does it relate to other sources?
- Do any graphics convey relevant information? Are sources cited?

- **Is the content sound?** Compare the website's content with what you find in your textbook, other assigned course readings, and other material available to you on the library shelves. You will place yourself in a good position to be critical of what you read online if you first read your textbook and other reading assignments.

- **Is the content biased?** Who wrote the information? Does the information reflect the special interests of a sponsor? Does the material support a single stance, political or otherwise? You can get a sense of balance—and thus better appreciate the significance of any bias—if you look for sites offering contrasting views.

- **Is there an ideological bent to the material presented?** Knowing the source of the material, and its sponsors, may assist you in identifying whether or not the site is intended to get readers to accept a particular line of argument.

- **Can you verify the specific information presented on the website?** Don't accept everything at face value. Question it and, if possible, verify it against other sources using, for instance, library-based research.

- **Is documentation provided?** Are footnotes, endnotes, a list of references, or other forms of citations provided? Do they include print and Internet sources, or just Internet sources? How pertinent are they to the material presented? If references are included, is there a clear rationale for their inclusion? Are references fully cited? Just as you must carefully consider documentation in books and journals, you should be wary—indeed, very wary—of the kind and form of documentation provided on the Internet. The need for documentation will vary, depending on the source of the material: a governmental agency may not offer any documentation at all, whereas an individual who purports to be a scholar, and wants you to believe everything that he or she presents, may provide more supporting documentation than is strictly necessary. As in all critical evaluations of others' work, it's important to ask whether the documentation is germane and complete. Have the sources been used with care and accuracy? Can you think of any major sources that have been neglected?

- **Are the links relevant and appropriate for the site?** Are the links comprehensive or are they limited? The degree to which links are appropriate, as an expansion of the website's purpose, may give you a clue as to the worthiness of the site itself.

- **Should I be wary of open-source sites?** Yes! For example, *Wikipedia* has been in existence since 2000 as an online "encyclopedia" that "anyone can edit." The "visitors" who volunteer to write and edit "articles" do not need specialized qualifications to contribute: most of the entries can be edited by anyone with access to the Internet. Because of its open nature, where entries are based on what people choose to submit or amend (including problematic content), critics charge that *Wikipedia's* content lacks authority, accountability, is unreliable, and subject to vandalism (insertion of false information). Articles do not undergo the same formal, peer review processes as academic journal articles; *Wikipedia* warns users that not all of its information is accurate, unbiased, or comprehensive. Many academics and librarians consider that *Wikipedia* is unsuitable as a reference source; indeed, your instructors may discourage you from using citations from this source in your assignments.

- **Above all, be critical, not gullible!** Don't be fooled into accepting everything you find on the Net. This also means that you should not take the easy route of basing your research on something that is "there," just because it is.

Table 2.1   An overview for evaluating websites

Questions to Ask	Points to Check
A. Looking at the URL:	
1. Is this a personal page or site?	• check for ~ or % or *users, members, people*
2. What type of domain is this? Is the information on the site appropriate for the domain?	• .edu? .gc? .com? .org? .net? country? • other sources?

3. Who is publishing the page? Is it published by an entity that makes sense? Does it correspond to the name of the site? Have you heard of this entity previously?	• check the publisher or domain name • look for links such as "about us", "background", "who am I", "biography", "sponsors" • you can place greater reliance on information that is published by the source

B. Scanning the page perimeter:

1. Who, or what agency, wrote the page? Can you tell what values they stand for?	• look for name of author, organization, agency, or institution responsible for the page (someone who claims to be responsible for the content) • truncate back the URL (one "/" at a time) if you can't find this information (an email contact is not enough)
2. When was the page last updated? Is it current enough for your purposes?	• date usually is found at the bottom of a web page (there may be a date in the URL) • check dates on all pages on the site • do not use undated factual or statistical information
3. What are the author's credentials?	• check whether the author's purported background or education suggests qualifications sufficient to write on the topic • an email address with no additional information about the author is insufficient to assess credentials

C. Considering quality:

1. Are the author's sources well documented? Did the author get the information from published scholarly or academic journals and books?	• look for footnotes and "links", "additional sites", "related links", etc. • explore any footnotes or links that may refer to documentation; are the publications or sites reputable? reliable? scholarly? real? (on the Web it is possible to provide totally fake references)
2. Is the information complete? If the site provides second-hand information, is it reproduced in an unaltered state?	• you may need to find the original source to determine whether the copy is complete and not altered (if it's retyped, it could be altered easily) • a legitimate source from a reputable journal or other publication should be accompanied by the copyright statement and/or permission to reprint (be suspicious if it's not present)

3. Does the author provide links to other resources on the topic?
- are the links working? are they well chosen?
- are there links to other points of view? (links that offer opposing viewpoints are more likely to be balanced and unbiased than pages offering one view only; be sure to check for bias, particularly when you agree with something!)

D. Checking with others:

1. Who links to the page or site? Are there many or few links? What opinions are expressed?
- find out other web pages that link to the site: in Google you can search: link:all.or.part.of.url

2. Is the page listed in one or more reputable directories or pages?
- good directories list a small portion of websites, so inclusion in a directory is significant (but the directory may not be totally positive about the site)
- check http://lii.org, or http://infomine.ucr.edu, or http://about.com

3. What do others say about the author or responsible authoring body?
- look up the author's name in Google or Yahoo!
- see which blogs refer to the site and what they have to say about it (use site name, author, or URL to do this on Google Blog Search)

E. Stepping back and thinking:

1. Why was the page put on the Web? Is it suitable for your purpose?
- to inform? provide facts and data? explain? persuade? sell? entice? share or disclose? other reasons?
- be aware of the full range of possible intentions behind web pages

2. Is the tone of the site ironic? Is it satire or parody?
- it's easy to be fooled; be sensitive to the possibility of becoming a victim!

3. Is the web page as good (credible and useful) as the books, journals and other resources available in print or online through the library?
- generally, published information is considered to be more reliable than information found on the Web. However, a great many reputable agencies, institutions, societies, publishing houses, news sources, and governments make excellent material available through the Web—take the time to check it out carefully.

SOURCE: After The Regents of the University of California. (2008). Evaluating web pages: techniques to apply & questions to ask. http://www.lib.berkeley.edu/TeachingLib/Guides/Internet/Evaluate.html.

COPYING MATERIAL FROM THE INTERNET

You may elect to copy material directly from a website by highlighting something on a page, copying it, and then pasting it into a word processing document. If you do this, be sure to identify exactly what has been copied, with full and correct source and page information, and the date you accessed the site. Carelessness in this regard could lead you to inadvertently use the material without including quotation marks or without providing source information. The excuse of carelessness will not save you from a charge of plagiarism.

REMAIN OPEN TO "FRINGE" INFORMATION

While you want to be sure your information is sound, it's a good idea to remain open to Internet sources that provide access to "fringe" information. For instance, many newspapers that are neither "mainstream" nor officially "accepted" in their countries of origin are available online, and they may convey information that otherwise would never be released. Similarly, oppressed minorities may use the Internet to get their messages to an international audience. You should be cautious when using such sources, but the insights gained from them sometimes can be invaluable. As always, be sure to cite your sources (see Chapter 11).

SELECTED WEBSITES OF INTEREST TO GEOGRAPHERS AND ENVIRONMENTAL SCIENTISTS

Since millions of websites exist, determining which are most pertinent and useful to your studies can be time consuming. A list of recommended websites is provided in Appendix III (pp. 305–10). Many of the sites provide links to other sites, so be open to trying the links. Be sure to remember and apply the points raised above about how to check the validity of websites.

NOTES

1   For an excellent overarching exploration of the discipline of geography, see Geoffrey J. Martin, *All Possible Worlds: A History of Geographical Ideas*, 4th ed. (New York: Oxford University Press, 2005). For discussions of recent developments in human geography and various approaches to research see Stuart Aitken and Gill Valentine, *Approaches to Human Geography* (Sage Publications Inc., 2006); Ronald J. Johnston and J.D. Sidaway, *Geography and Geographers: Anglo-American Human Geography since 1945*, 6th ed. (London: Hodder Arnold,

2004); Arild Holt-Jensen, *Geography: History and Concepts—A Student's Guide*, 3rd ed. (London: Sage, 2000); and the many entries in Ronald J. Johnston, et al., eds., *The Dictionary of Human Geography*, 4th ed. (Oxford: Blackwell, 2000). Counterparts in physical geography include Kenneth J. Gregory, *The Changing Nature of Physical Geography* (London: Arnold, 2000), and David Thomas and Andrew Goudie, eds., *The Dictionary of Physical Geography*, 3rd ed. (Oxford: Blackwell, 2000). Thematic reviews of some sub-fields, with reference lists, appear regularly in *Progress in Human Geography*, *Progress in Physical Geography*, *Progress in Environmental Science*, and *Progress in Development Studies*. Other works which examine aspects of geography and associated concerns include Gary L. Gaile and Cort J. Willmott, eds., *Geography in America* (Columbus, Ohio: Merrill, 1989); Gary L. Gaile and Cort J. Willmott, eds., *Geography in America at the Dawn of the 21st Century* (New York: Oxford University Press, 2006); Ronald J. Johnston and Paul J. Cloke, eds., *Spaces of Geographical Thought* (London: Sage, 2005); Paul Cloke, et. al., *Practising Human Geography* (London: Sage, 2004); Noel Castree, Alasdair Rogers, and Douglas Sherman, eds., *Questioning Geography: Fundamental Debates* (Oxford: Blackwell, 2005); John A. Agnew, K. Mitchell, and G. Tuathail, *A Companion to Political Geography* (New York: Blackwell, 2003); James S. Duncan, Nuala C. Johnson, and Richard H. Schein, eds., *A Companion to Cultural Geography* (Oxford: Blackwell, 2004); Lise Nelson and Joni Seager, eds., *A Companion to Feminist Geography* (Oxford: Blackwell, 2005); Linda McDowell and J.P. Sharp, eds., *A Feminist Dictionary of Human Geography* (London: Arnold, 1999); Paul Cloke, P. Crang, and M. Goodwin, *Introducing Human Geographies*, 2nd ed. (London: Hodder Arnold, 2005); David Atkinson, et al., eds., *Cultural Geography: A Critical Dictionary of Key Concepts* (London: I.B. Tauris, 2005); Gary Bridge and Sophie Watson, eds., *A Companion to the City* (Oxford: Blackwell, 2002); J. Stillwell and G. Clarke, eds., *Applied GIS and Spatial Analysis* (Chichester: Wiley, 2004); Alexander B. Murphy and Douglas L. Johnson, eds., *Cultural Encounters with the Environment: Enduring and Evolving Geographic Themes* (Lanham, MD: Rowman & Littlefield, 2000); and David B. Knight, *Landscapes in Music: Place, Space and Time in the World's Great Music* (Lanham, MD: Rowman & Littlefield, 2006).

2   Donald Meinig, "The Life of Learning," in *Geographical Voices: Fourteen Autobiographical Essays*, Peter Gould and Forrest R. Pitts, eds. (Syracuse: University of Syracuse Press, 2002), p. 205.

3   In libraries—and academic discussions—scholarly journals are not called "magazines." The latter term is reserved for periodic (generally weekly or fortnightly) popular titles, such as *Time*, *Maclean's*, and *The Economist*. You will find that some professors will refer to a journal article as a paper.

4   See Appendix II for a list of many of the different types of journals of interest to geographers and other environmental scientists.

5   See the list of disciplinary and interdisciplinary dictionaries in footnote 2 of Chapter 1 (p. 18). Pertinent encyclopedias include: National Council for Science and the Environment and the Encyclopedia of Earth and the Environmental Information Coalition, *The Encyclopedia of Earth* (on-line: http://www.eoearth.org); John E. Oliver, ed., *Encyclopedia of World Climatology* (New York: Springer, 2005); David E. Alexander, et al., eds., *Encyclopedia of Environmental Science* (New York: Springer, 1999); Tim Forsyth, ed., *Encyclopedia of International Development* (London: Routledge, 2005); Barney L. Warf, ed., *Encyclopedia of Human Geography* (New York: Sage, 2006); Andrew Goudie, ed., *Encyclopedia of Geomorphology* (London: Routledge, 2004); Andrew Goudie and D.J. Cuff, eds., *Encyclopedia of Global Change: Environmental Change and Human Society* (Oxford: Oxford University Press, 2002); and R.E. Munn, et al., eds., *Encyclopedia of Global Environmental Change* (New York: Wiley, 2002). Also useful is N.J. Smelser and P.B. Baltes, eds., *International Encyclopedia of the Social and Behavioral Sciences* (New York: Elsevier Science, 2001).

6   For historically important figures see the series produced by the International Geographical Union, Study Group/Commission on the History of Geographical Thought, *Geographers: Biobibliographical Studies* (London: Mansell, 1977–) and books such as Harm J. de Blij, *Wartime Encounter with Geography* (Lewes, Sussex: The Book Guild, 2000); Yi-Fu Tuan, *Who Am I?: An Autobiography of Emotion, Mind, and Spirit* (Madison: University of Wisconsin Press, 1999); Peter Gould, *Becoming a Geographer* (Syracuse: Syracuse University Press, 1999); Anne Buttimer, *The Practice of Geography* (London: Longman, 1983); David Lowenthal, *George Perkins Marsh: Prophet of Conservation* (Seattle: University of Washington Press, 2000); Geoffrey J. Martin, *The Life and Thought of Isaiah Bowman* (Hamden, Conn.: Archon Books, 1980); Gary S. Dunbar, *A Biographical Dictionary of American Geography in the Twentieth Century* (Baton Rouge, La.: Geoscience Publications, Department of Geography and Anthropology, Louisiana State University, 1992); and Peter Gould and Forrest R. Pitts, eds., *Geographical Voices: Fourteen Autobiographical Essays* (Syracuse: Syracuse University Press, 2002).

7   Based on Barbara Farrell, "Map Evaluation," in *World Mapping Today*, B. Parry and C.R. Perkins, eds. (London: Butterworth, 1987), pp. 27–34, by permission of the author.

8   The *Library of Congress Subject Headings* volumes ("the big red books") are available in the Reference section of the library. Check with a librarian if you can't find them. How can they be of use? If you are doing research on population

issues in China, you might start with broad keywords, such as "China and population." But be prepared to reconsider your keywords if the results are not discriminating enough. For example, using the keywords noted above, a recent Google search identified 31,900,000 results, all within 0.28 seconds! A second search was made, using the Library of Congress subject identifier "China–population": "only" 8,990,000 sites were identified. A more targeted search engine, Google Scholar, was then used. It found 1,150,000 sites using "China and population," and 27,000 sites using "China–population." Obviously, further refinement would be necessary for any research on this general topic. It is important to try various words, and combinations of words, when using any search engine.

CHAPTER 3

Writing and Reading Lecture Notes

The notes you take in lectures or that you jot down while reading your textbook are for you—not anyone else—to use when you study for examinations, prepare to participate in a class, or write papers. Good notes don't just happen—you need to be systematic in your approach so that you can use them to the fullest possible benefit at a later time. This chapter briefly explains why notes are important and suggests some ways to improve your note-taking skills.

WRITING HELPS YOU REMEMBER

As you listen to a lecture, watch an instructional movie, participate in a class discussion, work on a lab experiment, or read a book or article, keep in mind that you will remember the points raised much more clearly if you take notes. Don't just write them for the moment—write them for the future, whether that future is two weeks, two months, or even two years away.

READING NOTES

Finding your research material in books and journals is one thing; taking notes that are dependable and easy to use is another. Good, concise notes taken as you read the material can be invaluable. Writing them helps you concentrate while you read, and the act of writing itself helps you remember what you have read. With time you will develop your own best method, but for a start you might try recording each new idea or piece of evidence on a separate index card or sheet of paper; the number you need will depend on the range and type of your research. If you choose to use a laptop for note taking, think very carefully about your file management plan. You may decide to start a new file for each reading. If so, be sure to name each file with relevant titles so that you can easily organize the material. Be sure to keep all the files together in one folder, separate from any other files for different courses. The following suggestions will help you make the most of your reading notes.

- Use an easily recognizable file name, if using a laptop.
- Use index cards or single sheets of paper, if writing notes by hand.
- Start by meticulously recording the full particulars of the work you are reading: author and title information, place of publication, publisher, year of publication, total pages of the work (preliminary pages, which usually are in lower-case Roman numerals [e.g., i–xxv], and the text pages, which will be in Arabic numerals [e.g., 1–25]), and, if a journal, the volume and issue numbers. Also record the library catalogue number, in case you want to find the work again. Accurate recording of all the data may save you a last-minute dash to the library to double-check a source just before completing an assignment. Nothing is more frustrating than using a piece of information in an essay only to find (perhaps under a severe time constraint) that you aren't sure where it came from, or that you have neglected to copy down all the publishing information.
- At the top of each new page, card, or file make some shorthand reference to the source so that you can later link it to the full bibliographic information.
- Don't mix quotations or paraphrases from several sources on the same card or sheet.
- If you take several ideas from one source, put the main bibliographic details about the author and work on one card or page, and then use a separate card or page for each particular idea or theory. That way you will be able to shuffle the material later, to help you better organize your thoughts and to put them in the order in which you will use them when you write.
- Concentrate and be selective. Not everything in a chapter or article will be equally important, so try to identify the points that are most relevant to the particular assignment and, more generally, the course for which you are reading the material.
- Read critically. Give careful consideration to the author's argument and try to separate opinions from facts.
- Writing out long passages from your readings may not be the best use of your time. Rather, read a section and then make concise notes that highlight the main points. When completed, your notes will reflect the outline of the work.
- Sometimes it's helpful to underline or highlight particularly important quotations or paraphrases in your notes, or to make photo-

copies of longer passages. When you do underline, include a note to remind yourself that the underlining was not in the original. *Never underline library copies of books and journals.* And remember that neither underlining nor photocopying will help you to think in the way that writing your own notes does.

- Use headings that will remind you of details; these may differ from the ones used by the author. It can also be useful to record both your own suggestive headings and those of the author.
- Be sure to write down the page numbers as you make notes. This is important when you are paraphrasing from a source, but it becomes vital when you write down a direct quotation. When a passage runs on for more than one page, it's a good idea to put a short horizontal line in the left margin with the first page number above and the second below; an additional mark within the sentence will indicate exactly where the page break was. Knowing the exact page source will be important if you end up citing only part of the quotation.
- If you are using a laptop, the same pointers just identified apply (see sample below). Note the use of a line in the left margin and // to show the location of the page break in the quotation. Be sure to identify your own comments, as suggested (within brackets, *in italics*) in the following sample entry.

Simmons, I.G., *Changing the Face of the Earth* (Oxford: Blackwell, 1990).

Discussion of "woods and forest" (pp. 161–173). In section on erosion (pp. 166–167):

(166)   "Soil loss from open areas like cropland and landslides is
——   2.3 times that from forests in the Hima//laya: cropped areas
(167)   may lose 64 kg/ha/yr whereas dense forest yielded only
       25 kg/ha/yr."

[*This may be useful in the essay on.* . . .]

- If you decide to copy excerpts from a work, be sure to record the location of any omissions. For instance, if you start a quote in the middle of a sentence, indicate the omission of the earlier material with an *ellipsis: three* or, when appropriate, *four* dots or periods (see p. 250).

- Check that quotations are copied correctly, word for word, as in the original. If there is a spelling error or a doubtful word or expression, it is customary to put [*sic*] (Latin for "thus," in square brackets) immediately after the item, to indicate that your citation is faithful to the original, error and all!

- Record clearly whether your notes are direct quotations, paraphrases, or your own comments. For exact quotations, use quotation marks. If you paraphrase, use a special mark at the beginning and end for later reference. You don't want to have your instructor point out that you must have "quoted" your own paraphrase!

- If you use abbreviations, be sure you know what they stand for. If you develop your own scheme for abbreviating, keep a record that you will be able to find in the future, in case you forget what any of your abbreviations mean.

- After you have read a work, write down your personal reactions to it. This record could be useful later if you are asked to comment on the relative worth—for you—of the various assignments, or if you need to write a review or research paper requiring evaluative comments. This record of your reactions also will be helpful at the review stage—by which time you may have developed different insights into a work as a result of other things you have read or heard since the first reading. If that is the case, jot down the new thoughts as well. Later, when you look at these varied insights, you will be able to see a progression in your thoughts. These notes may be of use if you are asked to provide your evaluation and comments on the usefulness of a source in an annotated bibliography.

- If you have handwritten your notes, staple or clip together the cards or pages that pertain to one reading—otherwise you may find half your notes on a particular article written on one card or sheet and notes taken on something entirely different on the next!

- The same organizational care needs to be taken when you use your laptop. Save all of your reading notes on a particular topic in the same folder, so you don't have to search for them in several different folders. You may wish to use separate files for each reading. Files can be merged and reorganized once you have completed the readings for an assignment. Print copies of the reading notes for ready reference for when studying or writing an assignment.

The reading log

Keep a log of everything you read for each course. You will be pleasantly surprised by how easily a glance at a log will trigger memories of what you have
read, even before you turn to your more detailed notes. The log should include
full referencing information, so that you will have a ready-made list of documented sources on selected course themes and topics. Such a list can be
invaluable when compiling bibliographies for other purposes. You can keep
your log in a notebook or on cards. Cards are more flexible, since they can be
shuffled in many ways for different course needs, but a notebook log has the
benefit of keeping everything read for any particular course together in one
place. You might also choose to computerize your log. If you do so, be sure
to keep a back-up copy in case you accidentally erase part—or all—of the file
while transferring the information for use in another project.

LECTURE NOTES

Lecture notes—and notes written during other types of classes—can be
invaluable. Writing them is a serious business because so much depends on
them. The following suggestions will help you write, and thereafter use, good
notes.[1]

   You may be permitted to take your laptop to the lecture room, or you may
be allowed to use a laptop during smaller classes, including discussion groups
and seminars. Remember, though, that the use of laptops in small class settings can be distracting, to you and others, and as a result some professors
may not allow their use. The comments below pertain to both forms of note
taking—written on paper and typed on a laptop.

Preparation
- Read the assigned pre-lecture readings. Taking notes when you read
  the assigned material will help you provide a framework within
  which you can place the lecture. An additional benefit to reading
  your assignments before the lecture is that you will be more attentive to any information given in class that does not appear in writings you have been assigned to read.
- If you use a laptop, have a separate folder for each course. Within
  the folder for each course, have separate files for each lecture. Be sure
  to title the files clearly so you can go directly to the one you want at
  a later time.

- In a sense, begin writing lecture notes *before* the lecture starts. The topic of the day may be listed on the instructor's course outline—although often the latter is only a guide, not a guarantee. Having read the assigned material, you will be in a position to know what the lecture topic of the day is likely to be. In other words, you will enter the lecture room with your mind ready for what is to transpire.
- Keep your hand-written notes, or the printed copies of your laptop note taking, together. Keep a separate set of notes for each course, either in a large binder with separators or in individual smaller binders, one for each course. A three-ringed or pressure binder is more useful than a fixed-page stapled or spiral-bound notebook, since it allows you to insert any handouts alongside your notes.

Active listening

It is critically important that you be an active listener. Don't be passive. Pay attention, take good notes, and remain attentive to anything new or different that is being presented. Two important points, as a start:

- **Attend lectures!** Don't rely on a friend to take notes for you. Since you may respond differently to the lecture material than your friend, you should be there to listen, write, and think for yourself about what is being said. Even if the instructor provides electronic copies of lecture materials on a website, before or after lectures, these do not replace your presence in the classroom.
- **Listening is an intentional and purposeful act.** Listening is different from hearing. How can you train yourself to listen? Read the assignments in advance and memorize definitions and disciplinary jargon. Go to lectures with the attitude that you intend to learn something new. Think positively, even if you find the lecturer sometimes is a bit boring. Focus on what the person is saying, not on how it is being said. Pay attention so you don't miss any essential information or genuine "nuggets" that may slip by if you are daydreaming. To help you to stay alert, occasionally move your body, especially your legs and feet, and periodically take some deep breaths. You will find that writing notes will help you to concentrate. The key is to truly listen so you can best understand what the instructor is saying.

The lecturer's intention and your responsibility

Remember that the lecturer's intention is not to summarize the textbook or other assigned readings, but to present some essential points about the topics under consideration. The lecturer may expand on one or more points in the textbook, compare and contrast different sets of material, give examples gained from field or lab research that has not yet been published, outline a concept or principle, work through a problem, present and discuss a hypothesis, explain particularly complex problems, synthesize seemingly disparate ideas, answer questions from the students, and so on. Your tasks are to listen and then link what you have heard with what you have read so that you can develop a comprehensive understanding of the key issues, appropriately grounded in the basic assumptions and methods of the discipline in question, and to reflect on and question the material.

Some Specific Pointers for Taking Notes

- Start your note-taking on the very first day of the course; important organizational information may be given that will be useful later.
- Begin the notes for each lecture on a clean sheet of paper or as a new file on your laptop. Record the date and lecture title.
- Pay attention to the organization of the lecture. Your instructor may provide an outline of the day's lecture, whether on the blackboard, by means of an overhead or a computer projection, or on a photocopied sheet. If so—and if the lecturer is faithful to the outline—it will be easy for you to jot down the major points, perhaps a key phrase or two, and any references provided. If there is no outline, or if the lecturer is disorganized, you will have to listen even more attentively; your outline will be especially helpful later, as you reflect on the purpose and content of the lecture.
- Don't spend the whole time writing. Stopping your note taking from time to time to just listen to the instructor can help your overall concentration.
- Listen carefully, think as you listen, and try to understand. Be attentive so that you can hear supporting information and insights as well as the main points.
- Write down definitions given by the instructor, since they may differ from the definitions of the same terms given in the course readings. If this is the case, consider the differences.

- Write down key words and phrases, names, places, dates, equations, and numbers.
- If you write down a quotation (given on the blackboard or screen, or read out), use quotation marks, and record the name of the author.
- Copy any sketch maps or diagrams that your instructor presents. (If you are using a computer, always make sure to have a few sheets of paper on hand for situations like this. Some, but not all, professors will let you download some of their figures from a course website.)
- If you use abbreviations for things said, be sure to record elsewhere what the abbreviations stand for so you know later what it is that you have written down!
- As the lecture proceeds, if a question pops into your mind, write it down on the left margin or within [brackets], to indicate that it is your thought. You can seek to answer it later, by yourself, or by speaking to the lecturer either at the end of the lecture or during the lecturer's posted office hours.
- Near the end of the lecture, be especially careful to listen for the speaker's conclusions; write them down accurately and concisely.
- If there is a question period after the lecture, listen closely; someone else's question may be just as important as the lecturer's response. Don't be afraid to pose a question, even if you think it's too simple— you may well find that others share your puzzlement and will be grateful to you for speaking up.
- Leave a space at the end of each day's notes so that you can add comments later, after you have reread and reflected on them in relation to other lectures and readings. This is easy when you use a laptop; however, be careful to distinguish your later comments (using *italics*, or perhaps a different colour of text) from those taken during the lecture.
- To be safe, if your notes are on the hard drive, save them to a backup CD, DVD, or memory device so you have two copies.

Three-part post-lecture reviews

Immediate post-lecture review

Review your day's notes as soon as possible after each lecture. Don't waste time rewriting your notes—unless they are such a mess that rewriting them is the only way to salvage them; however, some students like to type up their hand-written notes. If your notes are on a laptop, print them, so you can study from the hard copy. Don't study using just the laptop's screen; there is better retention when you use the printed page. Jot down any thoughts that come to mind that you wish to have clarified or expanded upon, by seeking assistance from the instructor, the TA, or doing your own reading.

PERIODIC REVIEW

Periodically review clusters of related lecture notes. If a logical break in themes is not evident, create your own. Do such reviews hand-in-hand with your reading notes and any other pertinent material, such as from the lab or field work. Since you will be taking several courses at once, stagger the periodic reviews so that you will have only one or two per week. By doing immediate post-lecture and periodic reviews you will develop a richer understanding of and appreciation for the course material.

THE FULL-COURSE REVIEW

This task, done just prior to the final examination, will be easier if you have systematically completed the two earlier types of review. If you have to do a last minute desperate "cram" it will be because you have failed to do the earlier forms of review.

Two warnings

First, if you miss a lecture, *don't* ask the instructor to lend you his or her lecture notes. The professor's lecture notes are personal notes and may not be in a form that will make sense to another reader, since they may just be guides to thought, to be developed as the lecture proceeds. Certainly they will not include comments made spontaneously that are often highly pertinent, nor will they include any questions raised by classmates. If you must borrow notes, get them from a classmate you trust to take good notes covering everything that has taken place in the class.

Second, instructors generally do not permit anyone to tape or otherwise record their lectures or seminar discussions. If you have a particular reason for doing so, as when you have a documented learning disability, you must obtain specific prior permission from your instructor, usually at the start of a semester. Many institutions have established specific procedures to facilitate such a request, so be sure to find out what you need to do before asking your instructor.

USING YOUR READING NOTES AND LECTURE NOTES

Reading notes and lecture notes will be useful for many purposes: when reviewing for tests, writing research essays, abstracts, research proposals, creating bibliographies (including annotated bibliographies), preparing for discussion groups, seminars, tutorials, labs, or field work, and, of course, studying for examinations. They also can be of use if you wish to formulate some key questions prior to seeing your instructor or TA in her or his office. Write the questions down so you remember them during such visits. (As a courtesy, you might give the instructor or TA a copy of your questions when you enter the office; this will facilitate the best use of the time available.) Because so much depends on the quality of your notes, you should take the task of writing them seriously. Doing a good job of note-taking right from the start will save you hours of frustration later.

NOTE

1   See also David B. Knight, "Getting the Best Out of Lectures and Classes," in *The Student's Companion to Geography*, 2nd ed., A. Rodgers and H.A. Viles, eds. (Oxford: Blackwell, 2003), pp. 173–9.

CHAPTER 4

Writing a Report on a Book or an Article

The term *book report* covers a variety of writing assignments, including an abstract or summary of the contents of a book or an article, an analytic report containing some evaluation, a sophisticated literary review, and an annotated bibliography. The following guidelines cover all four kinds of reports. Before you begin your assignment, be sure to check with your instructor to find out exactly which type of report is expected.

THE ABSTRACT OR SUMMARY

An abstract is a pithy summation of a work. Brevity and clarity are the keys to this kind of summary. If your instructor gives you a word limit, that limit will tell you exactly how many words you have in which to briefly, but accurately, sum up the content of the work in question. An abstract is not evaluative: it does not say anything about your reaction to the work. It simply records, as accurately as possible, in as few words as possible, your understanding of what the author has written.

How can you summarize the contents of a 20-page article, or a 300-page book? Given that the word count is vital in an abstract, you will have to be a strong editor. The following pointers may help.

1. **Determine the author's purpose.** An author writes an article or book for a reason: to cast some new light on a subject, to propose a new theory, or to bring together the existing knowledge in a particular field. Whatever the purpose, you have to discover it if you want to understand what guided the author's selection and arrangement of the material. The best way to gain insight into the author's intent is to check the table of contents, preface, and introduction of a book, or the introduction and headings of an article, and, generally, to skim the work in order to get a quick overview. The details will be much more understandable once you know where the work as a whole is going.

2. **Reread carefully and take notes.** A second, more thorough reading will be the basis of your note-taking. Since you already have determined the relative importance that the author gives to various ideas, you can be selective and avoid getting bogged down in a welter of unimportant detail. Just be sure that you don't neglect any crucial passages or controversial claims. When making notes condense the ideas. Don't take them down word for word, and don't simply paraphrase them. You will have a much firmer grasp of the material if you resist the temptation to quote. Force yourself to interpret and summarize. This approach will also help to make your report concise—remember, you want to be brief as well as clear. Condensing the material as you take notes will ensure that your report is a true summary, not just a string of quotations or paraphrases. You can start your summary in point form.

3. **Give the same relative emphasis to each area that the author does.** Do not simply list the topics in the work, or the conclusions reached: discriminate between primary ideas and secondary ones.

4. **Follow the book's order of presentation.** Strictly speaking, a simple summary need not do so, but it usually is safer to follow the author's lead. That way your summary will be a clear reflection of the original.

5. **Follow the chain of the arguments.** Do not leave any confusing holes. You won't be able to cover every detail, of course, but you must trace all the main arguments in such a way that they make logical sense.

6. **Mention (briefly) the key evidence supporting the author's arguments.** Without some supporting details, your reader will have no way of assessing the strength of the conclusions.

7. **Write your first draft.** Using complete sentences, write a comprehensive summary of the contents of the work.

8. **Editing, counting, re-editing, and re-counting.** Editing what you have written so that the sentences become tight, with no wasted words, is a difficult task, but with practice you soon will be able to reduce your abstract to the required number of words. If your sentences seem choppy or disconnected (because you are leaving so much of the original out) you can use linking words and phrases (see p. 221) to help create a

flow and give your writing a sense of logical development. After a first editing, don't be dismayed if the word count is still too high. Edit, and edit again, cutting down sentences while keeping all the main points in view. You will be surprised to see how complex ideas can be reduced to fewer and fewer words. In the end you will have a summary that clearly conveys the contents of the work, within the required word limit.

9. **Reread for accuracy, comprehensiveness, and mistakes.** Before handing the abstract in to your instructor, check it one more time. Are any major points missing? Have you recorded fairly what is in the work (not what you wish had been in it)? Are there any grammatical or spelling errors?

10. **Check for proper identification.** For a book, remember to check that you have included the name of the author or editor, the title of the work, and the full publication information. For a book chapter, include all of the above details plus page numbers. For a journal article, include all of the above details plus volume and issue numbers. See Chapter 11 for more information about how to document your sources. Remember, too, to include your own name and (if called for) student number, the course name and number, and date of submission.

THE ANALYTIC BOOK REPORT

An analytic book report—sometimes called a book review—not only summarizes the main ideas in a book but at the same time evaluates them. It's best to begin with an introduction, and then follow with the summary and evaluation. Publication details usually are identified at the beginning, but may be placed at the end also.

Writing guidelines

INTRODUCTION

You should provide all the background information necessary for a reader who is not familiar with the book. Here are some of the questions you might consider:

- What is the book about?
- What is the author's purpose? What kind of audience is he or she writing for? How is the topic limited? Is the central theme or argument explicitly stated, or only implied?

- What documentation is provided to support the central theme or argument? Is the documentation sound? Is it presented clearly, and does the discussion of it develop logically? Does the documentation support the author's contentions and conclusions?
- Do the title and subtitle fairly identify the book's content?
- How has the author divided the book into chapters? Do the chapter titles accurately reflect each chapter's contents?
- How does this book relate to others in the same specific field? To other works in the same broad area?
- What are the author's background and reputation? What else has he or she written?
- Are there any special circumstances connected with the writing of this book? For example, was it funded by or written for a scientific organization, a corporation, an ideologically-based think-tank, an advocacy group, or a governmental agency?
- What sources has the author used? Does the list of references (or the bibliography) include both historically important and recent works in the field?

Not all of these questions will be applicable to every book. Nevertheless, an introduction that answers at least some of them will make it easier for your reader to appreciate what you will later state by way of evaluation.

SUMMARY

Obviously you cannot analyze a book without discussing its contents: the basic steps are the same as those outlined above for the abstract or summary. You may choose to present a condensed version of the book's contents as a separate section, to be followed by your evaluation, or you may prefer to integrate the two, assessing the author's arguments as you present them. Again, remember that you are discussing someone else's work: review the book the author actually wrote—not the one you would like him or her to have written. In short, be fair.

EVALUATION

In evaluating the book, you might ask some of the following questions:

- What is the author's stated purpose?
- How is the book organized? Are the divisions useful? Does the author give short shrift to certain areas? Is anything left out?

- On what kinds of sources does the author rely?
- What is the nature of the evidence presented? Is the evidence reliable and up-to-date? Is it presented and analyzed in an appropriate manner? Is it distorted or misrepresented in any way? Does the evidence support the author's ideas? Could the same evidence be used to support a different case? Does the author leave out any important evidence that might weaken his or her case? Is the author's position convincing?
- Is there an inherent ideological basis to the line of argument?
- What kinds of assumptions does the author make in presenting the material? Are they stated or implied? Are they valid?
- Does the book accomplish what the author set out to do? Does the author's position change in the course of the book? Are there any contradictions or weak spots in the arguments? Does the author recognize those weaknesses or omissions? Remember that your job is not only to analyze the contents of the book, but also to indicate its strengths and weaknesses.
- Does the author agree or disagree with other writers who have dealt with the same material or problem? In what respects?
- Is the book written clearly and is it interesting to read? Is the writing repetitious? Too detailed? Not detailed enough? Is the style clear? Or is it plodding, or jargonish, or flippant?
- Does the book raise issues that need further exploration? Does it present any challenges or leave unfinished business for the author or other scholars to pick up?
- To what extent would you recommend this book? What effect has it had on you?
- Finally, don't forget to comment on the essential "trappings." If the book has an index, how good is it? Are there illustrations? Are they useful? Is there a list of "further readings," or a full-fledged bibliography?

THE LITERARY REVIEW

The literary review is a variation of the analytic book review. The term "literary" refers to the presentation rather than to the material discussed: the review should stand on its own merits as an attractive piece of writing.

The advantage of a literary review is the freedom it allows you in both content and presentation. You may emphasize any aspect you like, as long as you leave your reader with a basic understanding of what the book is about. In most cases, your purpose is simply to provide a graceful introduction to the work based on a personal assessment of its most intriguing—or annoying—features. Just don't be too personal: some reviewers end up telling us more about themselves than about the book. Although a literary review is usually less comprehensive than an analytic report, it should always be thoughtful, and your judgment must never be superficial.

The best way to learn how to write a good literary review is to read some of them. Check the book review sections of magazines such as *The New Yorker*, the weekend editions of the *Globe and Mail*, *New York Times*, and *The Times* of London, or papers that specialize in reviews, such as the *New York Review of Books* and *The Times Literary Supplement*, to see different approaches. In addition to reviews of specific books, some disciplinary journals also publish overviews of the literature on a given subject, in which several books are examined; journals such as *Progress in Human Geography*, *Progress in Physical Geography*, and, on occasion, the *Annals of the Association of American Geographers* include such reviews. Pay particular attention to the various techniques that reviewers use to catch the reader's interest and hold it. The basic rule is to reinforce your comments with specific details from the book or books in question; concrete examples will add authenticity and life to your review.

ANNOTATED BIBLIOGRAPHY

An annotated bibliography is simply a bibliographic listing of works that includes a very short summary statement about each work. The statement normally is no more than a sentence or two in length. Think carefully about the central purpose of the book and its contents and then sum them up as briefly as possible. Generally, also, there will be a brief comment on the relative worth of the publication for the topic you are writing about. For example:

> Fong, Y.T., *Population Growth and Urban Development in China* (Hong Kong: Mei Ling Settlement Research Centre, 2006).
> Using census data and population estimates, Fong explores China's changing population from 1951 to 2001 and identifies the consequences for mid-sized cities. The burgeoning populations in such cities have created a strong demand for new housing stock. This general survey provides a context for my focus on housing needs in the city of Yiphee.

CHAPTER 5

Writing an Essay

If you dread writing an academic essay, you will find that following a few simple planning and organizational steps will make the task easier—and the result better.

THE PLANNING STAGE

Some students claim they can write essays without any planning at all. On the rare occasions when they succeed, their writing usually is not as spontaneous as it seems: in fact, they have thought or talked a good deal about the subject in advance, and come to the task with some ready-made ideas. More often, trying to write a lengthy essay without planning just leads them to frustration. They get stuck in the middle and don't know how to finish, or suddenly realize that they're rambling off in all directions.

By contrast, most writers say that the planning, or pre-writing, stage is the most important part of the whole process. Certainly the evidence shows that poor planning usually leads to disorganized writing; in the majority of student essays the single greatest improvement would not be better research or better grammar, but better organization.

The insistence on planning doesn't rule out exploratory writing (see p. 79). Many people find that the act of writing is the best way to generate ideas or overcome writer's block; the hard decisions about organization come after they've put something down on the page. Whether you organize before or after you begin to write, however, at some point you need to plan.

Know your source material

You don't write an essay on "something" out of thin air. You first need to become fully acquainted with the source material you plan to use. Before you start writing, finish analyzing your primary material, if it is from lab or field research, or from the results of statistically manipulated and tested information from a data set. If your writing is to be based on books and journal articles, first develop good reading notes so you can draw upon them in a well-organized manner when writing your paper (see Chapter 4). Be aware that some instructors prefer, or will demand, that you use materials other than textbooks. Get to know what is expected of you, and approach your writing accordingly, with appropriate use of sources.

Ask questions

Your essay may be on a topic for which you have done no prior research. Some instructors ask students to choose their topics from a list of suggestions; others simply identify broad subject areas and permit students to develop their own specific topics within them. In the latter case, since a subject area is bound to be too broad for an essay topic, you will have to analyze it in order to find a way of limiting it. The best way of analyzing is to ask questions that will lead to useful answers.

Chapter 2 outlined a formula of six linked questions: *what? where? when? why? how?* and *who?* Posing these five or six questions—"who?" is not relevant in some research areas—will lead you to five or six initial answers that will in turn lead to other questions. Most initial questions, and the answers to them, tend to be too general, but they will stimulate more specific questions that will help you to refine your topic and formulate *the* basic question. The most important question can be posed within your statement of purpose, or transformed into a thesis statement, perhaps as a hypothesis. Remember to make your statement limited, unified, and exact (see pp. 28–9).

Try the three-C approach

A systematic scheme for analyzing your subject is the three-C approach. It forces you to look at your topic from three different perspectives, asking basic questions about *components*, *change*, and *context*:

COMPONENTS
- What parts or categories can you use to break down the subject?
- Can the main divisions be subdivided?

CHANGE
- What features have changed?
- What temporal and spatial patterns can be observed?
- Is there a trend?
- What caused the change?
- What are the results of the change?

CONTEXT
- What is the larger issue surrounding the subject?
- In what tradition or school of thought does the subject belong?
- How is the subject similar to, and different from, related subjects?

WHAT ARE THE *COMPONENTS* OF THE SUBJECT?

In other words, how might your topic be broken down into smaller elements? This question forces you to take a close look at the subject and helps you avoid oversimplification. Suppose that your assignment is to discuss the implications for a downtown area related to the construction of a new shopping mall in the suburbs, at an intersection of a major highway that bypasses the town. You might decide that the basic components of the study are (1) the site and situational characteristics of both the downtown and the mall; (2) the types and sizes of stores attracted to the mall versus the types in the downtown; (3) the responses of shoppers; and (4) the responses of retailers in the downtown. If these components seem too broad, you may break them down further. For example, you might wish to sample specific types of stores, such as clothing stores, or perhaps just women's clothing stores.

Similarly, if you are analyzing the landscape imagery in a work of literature, you could ask, "What kinds of images does the author use?" (e.g., similes, metaphors, and allusions). Or you could ask, "What are the content groupings?" (e.g., urban versus rural images, or images of summer versus winter, etc.).

WHAT FEATURES OF THE SUBJECT REFLECT *CHANGE*?

In all research involving the environment, no matter what the disciplinary focus, there is a concern with process, or change from one state to another: how the physical and biological environments came to be as they are and how they continue to change; how varying beliefs, attitudes, values, and actions on the part of humans have led to the creation of different cultural landscapes (that is, landscapes as moulded and modified by human action); and how various future actions could lead to different consequences for the environment.

To return to the mall example, there are several questions we can pose regarding change. What level of governmental authority permitted this change (i.e., the creation of the suburban mall)? How did downtown retailing land-use patterns change as a result of the mall's creation? Over what period of time did they change? How did shoppers' spatial behaviour change? How did these changes manifest themselves in people's activities? Did these changes affect downtown merchants? How did those merchants respond? Did the merchants propose changes to the downtown to attract shoppers back from the suburban mall? Did the town government support those changes? What were the results?

WHAT IS THE *CONTEXT* OF THIS SUBJECT?

Into what particular school of thought or tradition does your research fit? What perspective will you take as you develop your ideas and interpretations—logical positivist? structuralist? humanist? feminist? realist? postmodernist? critical geography? Be aware of the context that your instructor wants you to work within.

Consider the mall example again. The mall itself is a site for study, not a context. The context is different for different researchers. A biologist may be concerned with the impact of the mall on local flora and fauna; a landscape architect may be interested in the form and function of the mall; a geomorphologist may investigate the impact of the mall's expansive sealed parking lot on surface or subsurface water flows; a logical positivist geographer may advance and test a hypothesis based on location theory, using data on shopping patterns, movements of people, the value of goods sold, land values, etc.; a structuralist may be concerned with the ways in which politics influenced the location of the mall; a humanistic geographer may be interested in the importance of symbolism in the mall as opposed to the downtown area, or in what people feel and think about the two locales; a feminist geographer may be concerned with the societal consequences of the gender distribution of jobs created in the mall and with the lived experiences of those who work or shop in the mall; a critical geographer will want to examine how the relations of power have influenced the way the locations have developed, with a concern about social exploitation; and so on. Whatever the context, it will have to be buttressed by citation and discussion of pertinent literature.

In short, the context for your topic will lead you to pose different questions, which in turn will lead to different types of insights. And remember, no single context (and associated approach) is more "correct" than any other! Just be aware that, since there are differences in underlying assumptions and methods, it is essential to determine a clear context for each project you tackle.

Analyzing a prescribed topic

Even if the topic of your essay is supplied by your instructor, you will need to analyze it carefully. Try underlining key words to make sure that you don't neglect anything. Distinguish the main focus from subordinate concerns. A common error in dealing with prescribed topics is to emphasize one portion while giving short shrift to another. Give each part its proper due—and make sure that you actually do what the instructions tell you to do. Note that to *discuss* is not the same as to *evaluate* or *trace*. These verbs tell you how to approach the topic; don't confuse them. The following definitions will help you decide how to approach your assignment:

outline   state simply, without much development of each point (unless asked).

trace   review by looking back—on stages or steps in a process, or on causes of an occurrence.

explain   show how or why something happens.

discuss   examine or analyze in an orderly way; this instruction allows you considerable freedom, as long as you take into account contrary evidence or ideas.

compare   examine differences as well as similarities (see pp. 81–2).

evaluate   analyze strengths and weaknesses, to arrive at an overall assessment of worth.

Creating an outline

Individual writers differ in their need for a formal plan. Some say that they never have an outline, and others maintain they can't write without one. Some can't write until they have a title, headings, and subheadings, whereas others decide these details only once most of the work is completed. Most writers probably fall somewhere in the middle. Whatever the details of your study, it will be helpful to jot down the main points in outline form. Even if you later find that you have to change the order because the development of your argument has taken you in directions you did not plan, it's important to start out with a reasonable idea of where you are headed. If you have special problems with organizing material, your outline should be formal and in complete sentences. On the other hand, if your mind is naturally logical, you may find it is enough just to jot down a few words on a scrap of paper. For most students, an informal but well-organized outline in point form is the most useful model. (The example below uses the *decimal system* for organizing the sections. For this and other ways to identify the sections of your essay, see Chapter 10). The headings and subheadings in an outline should be precise enough to give a good idea of what the final work will include. For example:

HYPOTHESIS: Changes in Botswana's political systems led to structural changes in the settlement system and to changes in use of resources.

1. Pre-colonial political and settlement systems
    1.1 Tribal structures and the power of chiefs
        1.11 Social systems and the all-encompassing nature of chiefs' powers
        1.12 A fluid socio-political system .
    1.2 Settlement system and the locus of power
        1.21 Environmental characteristics
        1.22 Three-part settlement system: lands, cattle posts, and villages
        1.23 Chiefs' powers and the seasonal utilization of resources

2. Consequences of colonial intrusion
    2.1 Freezing of a fluid socio-political system
        2.11 Colonial indirect rule
        2.12 Changes in the role of chiefs
    2.2 Societal changes
        2.21 Imposition of the "hut tax" and the need for paid labour
        2.22 Labour migrations to South Africa
        2.23 Changes in gender roles in resource use and development
    2.3 Changes to the settlement system
        2.31 The impact of absentee labourers on pastoral and agricultural activities
        2.32 Incipient break-up of the three-part settlement system

3. Independence and rapid change
    3.1 Tensions between old and new systems
        3.11 Chiefs and tradition versus elected "new men"
        3.12 "Modernization" of society
    3.2 Changed settlement systems and resource use
        3.21 Imposition of a modern settlement system and an extractive resource infrastructure
        3.22 Replacement of customary land tenure with the Tribal Grazing Land Policy and the privatization of land
        3.23 Impact on the settlement system of new laws governing use and development of resources

4. Conclusion

Appendices (if any; see pp. 168–9).

References or bibliography (see Chapter 11).

The guidelines for this kind of outline are simple:

- **Code your categories.** Use different sets of markings, such as the decimal system used in the example above, to establish the relative importance of your entries. Your word processing program likely will have a "numbering and bullets" insert capacity, with options within each category. Select what is best for your work. However, see the next point.

- **Categorize according to importance.** Make sure that only items of equal value are put in equivalent categories. Give major points more weight than minor ones.

- **Check lines of connection.** Make sure that each of the main categories is directly linked to the central thesis; then see that each subcategory is directly linked to the larger category that contains it. Checking these lines of connection is the best way to prevent essay muddle.

- **Be consistent.** In arranging your points, be consistent. You can move from the most important point to the least important, or vice versa, as long as you follow the same order every time.

- **Use parallel wording.** Phrasing each entry in a similar way will make it easier to be consistent in your presentation.

- **Be logical.** In addition to checking for lines of connection and organizational consistency, ensure that the overall development of your work is logical. Does each heading/idea/set of data/discussion flow into the next, leading the reader through your material in the most logical manner?

A final comment: be prepared to change your outline at any time in the writing process. Your initial outline is not meant to put an iron clamp on your thinking, but to relieve anxiety about where you're headed. A careful outline prevents frustration and dead ends—that "I'm stuck, where can I go from here?" feeling. But since the very act of writing usually generates new ideas, you should be ready to modify your original plan. Just remember that any new outline must have the consistency and clear connections required for a unified essay.

THE WRITING STAGE

Writing the first draft

Rather than labour for excellence from scratch, most writers find it easier to write the first draft as quickly as possible and then do extensive revisions later. However you begin, you can't expect the first draft to be the final copy. Skilled writers know that revising is a necessary part of the writing process, and that the care taken with revisions makes the difference between a mediocre essay and a good one.

You don't need to write all parts of the essay in the same order in which they are to appear in the final copy. In fact, many students find the introduction the hardest part to write. If you face the first blank page with a growing sense of paralysis, try leaving the introduction until later, and start with the first idea in your outline. If you feel so intimidated that you haven't been able to draw up an outline, you might try John Trimble's approach and charge right ahead with any kind of beginning—even a simple "My first thoughts on the subject are. . . ."[1] Instead of sharpening pencils or running out for a snack, try to get going. Don't worry about grammar or wording; scratch out pages or throw them away if you must—just start! Perhaps you can write down a key quotation from an important source and then jot down your initial thoughts on its importance. Remember, the object is to get your writing juices flowing.

Of course, you can't expect this kind of exploratory writing to resemble the first draft that follows an outline. You will need to do a great deal of changing and reorganizing, but at least you will have the relief of seeing words on a page to work with. Many experienced writers—and not only those with writer's block—find this the most productive way to proceed.

Developing your ideas: Some common patterns

The way you develop your ideas depends on your essay topic, and topics can vary enormously. Even so, most essays follow one or another of a handful of basic organizational patterns. Here's how to use each pattern effectively.

DEFINING

Sometimes a whole essay is an extended definition, explaining the meaning of a term that is complicated, controversial, or simply important to your field of study: for example, *environment* in biology, geography, and environmental studies; *monetarism* in economics; or *existentialism* in philosophy. More often, you will begin a detailed discussion of a topic by defining a key term, and then shift to a different organizational pattern. In this case, make your definition exact. It should be broad enough to include all the things that belong in the category and at the same time narrow enough to exclude things that don't belong. A good definition builds a kind of verbal fence around a word, herding together all the essential ideas and cutting off all outsiders.

For any discussion of a term that goes beyond a bare definition, it's important to give concrete illustrations or examples; depending on the nature of your essay, these could vary in length from one or two sentences to several paragraphs or even pages. If you are defining *environment*, for instance, you would probably want to discuss at some length the theories of leading geographers, ecologists, and others, since different approaches lead to different definitions. It may be helpful to consult specialized disciplinary dictionaries and encyclopedias (see note 2 in Chapter 1 and note 5 in Chapter 2), along with other pertinent literature.

In an extended definition it's also useful to point out the differences between the term in question and other terms that may be connected or confused with it. For instance, if you are defining *environment* you will want to distinguish it from *landscape*; if you are defining *place* you will want to differentiate it from *space*; if you are defining *climate* you will want to distinguish it from *weather.*

CLASSIFYING

Classifying means dividing something into parts according to a given principle of selection. The principle or criterion may vary; you could classify crops, for example, according to how they reproduce (e.g., tuberous *versus* seed reproduction), what soils they grow in, how long they take to mature, or what climatic conditions they require; members of a particular population might be classified according to age group, occupation, income, and so on. If you are using a system of classification, remember the following:

- All members of a class must be accounted for. If any are left over, you need to alter or add more categories.
- Categories can be divided into subcategories. You should consider using subcategories if there are significant differences within a category. If, for instance, you are classifying the workforce according to occupation, you might want to create subcategories according to income level or ethnic origin.
- Any subcategory should contain at least two items; otherwise it is simply a unique entry in your classification.

EXPLAINING A PROCESS

This kind of organization shows how something works or has worked, whether it be soil erosion, a rock slide, a weather cycle, a migration, or the stages in a military campaign. The important point is to work systematically, to break down the process into a series of stages. Although at times your order will vary, most often it will be chronological, in which case you should see that the sequence is accurate and easy to follow. Whatever the arrangement, generally you can make the process easier to follow if you start a new paragraph for each new stage. Be aware, of course, that any "stage" you identify may be an artificial highlighting within the overall process, i.e., unless your data reveal clear divisions. Take this into account as you write.

TRACING CAUSES

A causal analysis is really a particular kind of process discussion, in which certain events are shown to have led to or resulted from other events. Usually you are explaining why something happened. The main warning here is to avoid oversimplifying. If you're tracing causes, distinguish between a direct cause and a contributing cause, between what is a condition for something to happen and what is merely a correlation or coincidence. For example, if you find that in France both the age of the average driver and the number of accidents caused by drunk drivers are increasing, you cannot jump to the conclusion that older drivers are the cause of the increase in drunk-driving accidents. Similarly, you must be sure that the result you mention is a genuine product of a particular event or process.

COMPARING

People sometimes forget that comparing things means showing differences as well as similarities—even if the instructions do not say "Compare and contrast." The easiest method for comparison, though not always the best, is to discuss the first subject in the comparison thoroughly and then move on to the second:

Subject *X*: Point 1
              Point 2
              Point 3

Subject *Y*: Point 1
              Point 2
              Point 3

The problem with this kind of comparison is that it often sounds like two separate essays slapped together. To use it effectively, you must integrate the two subjects, first in your introduction (by putting them both in a single context) and again in your conclusion, where you should bring together the important points you have made about each. When discussing the second subject, try to refer repeatedly to your findings about the first subject ("unlike X, Y does such and such"). This method may be the wisest choice if the subjects for comparison seem so different that it is hard to create similar categories in which to discuss them—if the points you are making about X are of a different type than the points you are making about Y.

If it is possible to find similar criteria or categories for discussing the subjects, however, the comparison will be more effective if you organize it like this:

Category 1:   Subject X
              Subject Y

Category 2:   Subject X
              Subject Y

Category 3:   Subject X
              Subject Y

Because this kind of comparison is integrated more tightly, the reader can see more readily the similarities and differences between the subjects. As a result, the essay is likely to be more forceful. Similarly, if it is applicable to your task, you might first list "positive" consequences of, say, constructing a factory adjacent to a recreation park. You would follow that list with a list of "negative" consequences. By this means you'd be able to clearly identify the different potential outcomes and, from them, draw conclusions within the context of the purpose of the paper.

Introductions

The beginning of an essay has a dual purpose: to indicate both the topic and your approach to it, and to whet your reader's interest in what you have to say. An effective way to introduce a topic is to place it in a context—to supply a kind of backdrop that will put it in perspective. You step back a pace or two and discuss the area into which your topic fits, and then gradually lead into your specific field of discussion. Sheridan Baker calls this the funnel approach (see Figure 5.1, on p. 86).[2] For example, suppose your topic is the effect of acid rain on gravestones in a particular cemetery. You would want to open the paper with a general statement about acid rain, including a definition of that term, a brief discussion of key physical and chemical processes and the impact of human activities, and an overview of the spatial patterns of consequences. Only then would you turn to the examination of your case study, drawing on your field observations and measurements of corrosion on the gravestones. A funnel opening is applicable to almost any kind of essay.

It's a good idea to try to catch your reader's interest right from the start—you know from your own reading how a dull beginning can put you off. The fact that your instructor must read on anyway makes no difference. If a reader has to get through thirty or forty similar essays, it's all the more important for yours to stand out. A funnel opening isn't the only way to catch the reader's attention. Here are three of the most common leads:

- **The quotation.** Your reader's anticipation may be heightened if you include a short pertinent quotation, either as a stand-alone comment (within quotation marks, with the author identified) before you open your paper, or within your opening paragraph. This approach works especially well when the quotation is taken from the person or work you will be discussing.
- **The question.** Posing a rhetorical question will only annoy the reader if it's commonplace or the answer is obvious, but a thought-provoking question can make a strong opening. Just be sure that you answer the question in your essay.
- **The anecdote or telling fact.** This is the kind of concrete lead that journalists often use to grab their readers' attention. Save it for your least formal essays—and remember that the incident or fact must really highlight the ideas you are going to discuss.

Whatever your lead, it must relate to your topic: never sacrifice relevance for originality. Finally, whether your introduction is one paragraph or several, make sure that by the end of it your reader clearly knows the topic and direction you are taking.

Integrating quotations

A quotation can benefit an essay in two ways (see Chapter 11). First, it adds depth and credibility by showing that your position or idea has the support of an authority. Second, it provides stylistic variety and interest, especially if the quotation is colourful or eloquent. The trick to using quotations effectively is to make sure that they are germane to the topic and properly integrated with your discussion. You don't want them either to dominate your ideas, or simply seem to be tacked on, as an interesting side comment. To ensure that a quotation has the desired effect, always refer to the point you want the reader to take from it: don't let a quotation dangle on its own. Usually, the best way to do this is to make the point before the quotation:

> While doing field research in the Himalayas, McKinley had to walk for miles to reach the villages where he was to do the interviews. "It was a devilishly difficult task, carrying all of the gear, walking up the rocky valley floors, and crossing raging streams using very primitive rope bridges; but it was all very exciting."

Sometimes, however, a quotation can precede the explanation. This is a common approach to begin an essay, as noted earlier in this chapter. You must be sure, however, to explain the significance of the quotation so that your reader is not left wondering why you have included it.

If a complete sentence precedes the quotation, use a colon at the end of the introductory phrase; otherwise, use a comma:

> Simpson writes authoritatively: "The soils were wet."

> (or)

> Simpson stated her conclusion clearly, insisting that "The soils were wet."

If the end of the quotation comes at the end of a sentence, finish with a period—unless the sentence is a question, in which case finish with a question mark *after* the quotation marks:

> Could it be that Simpson was not correct in stating that "The soils were wet"?

The length of a quotation can vary from a short phrase woven into your sentence to a paragraph or more. If you take part of a sentence, be sure to use an *ellipsis* (see p. 250). If the quotation is four lines or longer, you will indent the quotation and not use quotation marks (see p. 256). Just remember, the longer the quotation, the greater the danger that it will overshadow rather than reinforce your own viewpoint. Don't quote any more than you really need.

Conclusions

Endings can be painful—sometimes for the reader as much as for the writer. Too often, the feeling that one ought to say something profound and memorable produces the kind of prose that suggests violins throbbing in the background. You know the sort of thing:

> The tremendous amount of soil erosion in the valley dramatically highlights the awful plight of the poor farmers, who for generations to come will suffer dreadfully from the loss of the very basis for their livelihood.

Why is this embarrassing? Because it is clearly phoney—no more than a grab-bag of clichés.

Experienced editors often say that many articles and essays would be better without their final paragraphs: in other words, when you've finished what you have to say, the only thing to do is stop. This advice may work for short essays, where you need to keep the central point firmly in the foreground and don't need to remind the reader of it. But for longer pieces, where you have developed a number of ideas or a complex line of argument, you should provide a sense of closure. Readers welcome an ending that helps to tie the ideas together; they don't like to feel they've been left dangling. And since the final impression is often the most lasting, it's in your interest to finish strongly. Simply restating your hypothesis or summarizing what you have done already is not forceful enough. What are the options?

- **The inverse funnel.** The simplest and most basic conclusion is one that restates your hypothesis *in different words* and then discusses its implications. Sheridan Baker calls this the inverse funnel approach, as opposed to the funnel approach of the opening paragraph.[3]

  One danger in moving to a wider perspective is that you may try to embrace too much. When a conclusion expands too far, it tends to lose focus and turn into an empty cliché, like the conclusion in the preceding example. It's always better to discuss specific implications than to leap into the thin air of vague generalities.
- **The new angle.** A variation of the basic inverse funnel approach is to reintroduce your argument with a new twist. Suggesting some fresh angle can add excitement to your ending. Beware of injecting an entirely new idea, though, or one that's only loosely connected to your original argument: the result could be jarring or even off-topic.

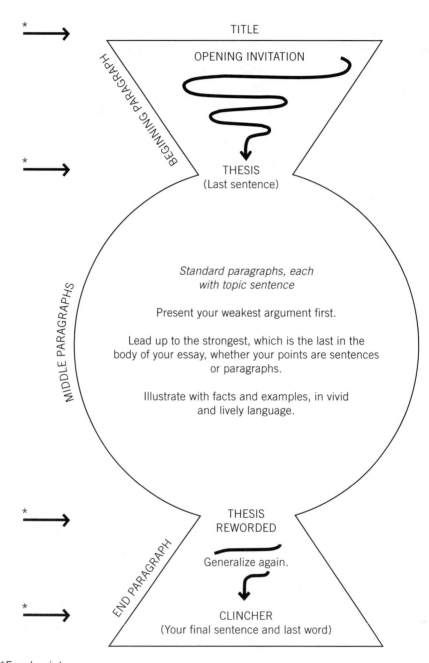

*Focal points

Figure 5.1   The funnel approach

SOURCE: Sheridan Baker and Laurence B. Gamache, *The Canadian Practical Stylist*, 4th ed. (Don Mills, ON: Addison-Wesley, 1998), p. 65. Reprinted by permission of Pearson Education Canada Inc.

- **The full circle.** If your introduction is based on an anecdote, a question, or a startling fact, you can complete the circle by referring to it again in relation to some of the insights revealed in the main body of your essay.
- **The stylistic flourish.** Some of the most successful conclusions end on a strong stylistic note. Try varying the sentence structure: if most of your sentences are long and complex, make the last one short and punchy. Sometimes you can dramatize your idea with a striking phrase or colourful image. When you are writing your essay, keep your eyes open and your ears tuned for fresh ways of putting things, and save the best for the end.

None of these approaches to endings is exclusive, of course. You may even find that several of them can be combined in a single essay. Whichever approach you take, just remember that a conclusion should never refer to facts that have not been cited already in the main body of your essay.

THE EDITING STAGE

Often the best writer in a class is not the one who can dash off a fluent first draft, but the one who is the best editor. To edit your work well you need to see it as the reader will; you have to distinguish between what you meant to say and what is actually on the page. For this reason it's a good idea to leave some time between drafts, so that when you begin to edit you will be looking at the writing afresh rather than reviewing it from memory. Without this distancing period you can become so involved that it's hard to see your paper objectively. Before you begin the serious task of editing, allow yourself to go to a movie or play some squash—do anything that will take your mind off your work. Also, if you are working on a word processor, remember that it can be hard to get a sense of the whole when you read your material as it appears on the screen, section-by-section. Print out a hard copy to edit on—and use a pencil.

Editing doesn't mean simply checking your work for errors in grammar or spelling. It means looking at the piece as a whole to see if the ideas are well organized, well documented, and expressed clearly.

Editing may mean adding some paragraphs, deleting others, and shifting still others around. Very likely editing means you will be adding, deleting, and shifting sentences and phrases. Experienced writers may be able to check several aspects of their work at the same time, but if you are inexperienced or in doubt about your writing, it's best to look at the organization of the ideas before you tackle sentence structure, diction, style, and documentation. And, don't forget to check the "little" things like subheadings, page numbers, numbers for illustrations (on the illustrations and in the text), spelling (and other details) in your illustrations, and the completeness and accuracy of your references. These items often are neglected—but instructors notice them.

What follows is a checklist of questions to ask yourself as you begin editing. Far from all-inclusive, the checklist focuses on the first step: examining the organization. You probably won't want to check through your work separately for each question: you can group some together and overlook others, depending on your own strengths and weaknesses as a writer.

An editing checklist

- Are the purpose and approach of my essay evident from the beginning?
- Are all sections of my paper relevant to the topic?
- Is the organization logical?
- Will the subheadings be meaningful to readers? Do they clearly identify the various sections in my work?
- Do my paragraph divisions give coherence to my ideas? Do I use them to cluster similar ideas and signal changes of idea?
- Do any parts of the essay seem disjointed? Should I add more transitional words or logical indicators to make the sequence of ideas easier to follow?
- Are the ideas sufficiently developed? Is there enough evidence, explanation, and illustration?
- Would an educated person who hasn't read the primary material understand everything I'm saying? Should I clarify some parts or add any explanatory material?
- In presenting my argument, do I take into account opposing arguments or evidence?
- Have I been accurate and fair in my representation of what my sources state?
- Have I cited all the sources I have used? Is the style of in-text citations consistent? (See Chapter 11.)
- Are my illustrations and tables useful? Do they present the data in the clearest, most effective way, or would a different form make a better presentation? (See Chapter 8.) Is my discussion of each one clear and complete?

- Are all my illustrations and tables numbered correctly? Have I identified them within the text?
- Are the tables complete and correct, with the source(s) noted?
- Have I checked my illustrations for correct spelling as well as for content? Have I included the source(s)?
- Do my conclusions accurately reflect my argument in the body of the work?
- Are the contents of my appendices useful? Are they complete and accurate?
- Is my reference list (or bibliography, whichever is appropriate) accurate and complete in all publication details?
- Is my title imaginative, informative, and precise?

Another approach would be to devise your own checklist based on the faults of previous assignments. This is particularly useful when you move from the overview to the close focus on sentence structure, diction, punctuation, spelling, and style. If you have a particular weak area—for example, irrelevant evidence, faulty logic, or run-on sentences—you should give it special attention. Keeping a personal checklist will save you from repeating the same mistakes over again.

As you edit, you will want to check for spelling errors. As noted in Chapter 1, your word-processing program may do this automatically, or you can initiate a separate check. Remember, however, that it won't identify words that are wrongly used (such as "there" when you mean "their"), and it may not catch numbers accidentally typed into words (e.g., swi5ft). Think of a computer spell-checker as a useful first check rather than a final one. Similarly, grammar-checkers, though considerably better than when they first appeared, are still not reliable. They will pick up common grammar errors and stylistic problems, but they do make mistakes and likely will never equal the judgment of a good editor.

Keep in mind, too, that for final editing, most good writers suggest working from the printed page, rather than from the computer screen. This point was made in Chapter 1, but it is worth repeating. In other words, print one draft before your final copy. Your eye will get a better sense of the text by perusing a whole page at a time than by scrolling down the screen line by line.

FORMATTING YOUR ESSAY

As discussed briefly in Chapter 1, a well-typed, proofread and corrected, attractive essay creates a receptive reader and, fairly or unfairly, often gets a higher mark than a sloppy paper (with bad grammar or atrocious spelling errors), which is more difficult to read. Good looks don't substitute for good thinking, but they will certainly enhance it! In other words, take care to prepare the cleanest copy of your work that you can before submitting it.

Always double-space your work and use margins of at least 2.5 centimetres (one inch) to frame the text in white space and allow room for your reader to write comments. Number each page at the top or bottom right-hand corner or at the top-centre of the page (without including a page number on the first page), and provide a neat, well-spaced cover page that includes the title of your paper, your name, the names of your instructor and the course, and a date. Make use of the formatting features such as bold and italics for appropriate emphasis, and choose suitable font faces. You will be safe if you use a serif font such as Times New Roman, usually 12-point pitch, for the body of the essay, and a sans serif font such as Arial for the title and headings.

Keep in mind, though, that while a computer can make your work look good, fancy graphics and a slick presentation won't replace intelligent thinking. Read over your work with a critical eye, in the knowledge that you can easily change something that is unsatisfactory. Remember that your computer is a tool for you to use—no more than that.

PROTECTING YOUR WORK

An essay involves a great deal of time and effort. You don't want to find that some or all of your material is "lost" due to a computer error, or one that you caused, such as by accidentally erasing a section of your work, or even a whole file. Lost files are the nightmare of the computer age. "Save" regularly! Keeping at least two copies of your work as you proceed, one on the hard drive and a back-up version on a CD, DVD, or memory device will give you peace of mind. You may be asked to submit a CD with your paper saved on it so it can be checked for plagiarism using special software. Keep the file at least until you have received the grade for the course. And always print an extra copy, for your files, in case the original is lost.

NOTES

1   John R. Trimble, *Writing with Style: Conversations on the Art of Writing* (Englewood Cliffs, NJ: Prentice-Hall, Inc., 1975), p. 11.
2   Sheridan Baker, *The Practical Stylist*, 5th ed. (New York: Harper and Row, 1981), pp. 24–5.
3   Ibid.

CHAPTER 6

Writing a Lab Report

Laboratory exercises have several purposes, but the most important is the "hands-on" working through of a problem, often using equipment or materials that cannot be used readily in the lecture room or at home. Whatever the specific content and purpose of the exercise, the overall goal is to clarify, amplify, and apply basic concepts, principles, theories, processes, and research methods in the particular field under consideration. Since labs generally work as building blocks, providing detailed insights into the subjects discussed in lectures and assigned readings, it's important that you complete the exercises systematically, in the order indicated by your instructor. See Appendix I for a list of pertinent weights, measures, and notation (pp. 300–1).

LAB BOOKS AND LAB ASSIGNMENT SHEETS

While some instructors give individual assignments for each lab, one by one, others will require you to buy a lab book. If you have a book, look through it to see what is coming later in the course; knowing what lies ahead can sometimes help you handle earlier assignments. In either case, before you begin any lab assignment, be sure you have read the required course material for that day and have thought about the relevant lectures. Whether or not you have a lab book, the following information will help you to complete your lab reports.

Purpose and reader

Since your lab report will be read either by your course instructor or a TA, you can assume that he or she is familiar with scientific terms; you may not have to define or explain them—unless you are specifically told to do so as part of the assignment. You can assume also that your reader will be on the look-out for any weaknesses in methodology or analysis and any omissions of important data. Usually you will be expected to give details of your calculations, but even when asked to provide just the results of your calculations, you should include any experimental uncertainty that might affect them.

Objectivity

Objectivity is critical in lab work. You must never allow your preconceived opinions or biases, or even expectations, to get in the way of the facts: otherwise you may, perhaps unwittingly, distort the findings. You must follow through with the experiment or exercise as objectively as possible and present the results in such a way that anyone who reads your report or attempts to duplicate your procedure would be likely to reach the same conclusions that you did.

Recording findings

All lab assignments involve making and recording observations (perhaps during an experiment) or measurements (reading a contour map, considering the application of an equation, determining grain size, measuring rates of erosion in a wave tank, etc.). Some assignments may require a numerical response. Others will call for words—perhaps nothing more than a simple "yes" or "no," "true" or "false." In other cases there may be fill-in questions, with space for only one or two words. Although you may be asked to write a little more, in general a precise, succinct summary statement or a point-form response will be all that you have time (or space) for. If a space is provided for your answer, make sure your writing is neither so small nor so large as to either "undersell" or "overkill"! In sum, read the question or direction and then consider your observation or response with care, answering in as few words as possible.

FORMAT

Since the information in scientific reports must be easy for the reader to find and assess, it should be grouped into separate sections, each with a heading. One of the differences between writing lab reports and writing essays is that in a report you should use headings and subheadings whenever possible, as well as graphs, tables, and diagrams. Most lab reports follow a standard order:

1. Title Page
2. Abstract (*or* Summary)
3. Introduction (*or* Statement of Purpose *or* Objective)
4. Experimental Design (*or* Methodology)
   (a) Materials
   (b) Method
5. Results
6. Discussion (*or* Analysis)

7. Conclusion
8. Attachments (*or* Appendices)
9. References

Some instructors will prefer that you identify *materials* within the *method* section, not separately. The guidelines presented below assume that the lab work to be reported is based on an experiment.

Title page
The first page of the report is your title page. It should include the title of the experiment, the date it was performed, your name, and the date of submission; for practical purposes it also should include the name and number of the course, the lab section number, and the name of your instructor and, if applicable, TA. The title should be brief but informative, clearly indicating the subject and scope of your experiment. Avoid meaningless phrases, such as "A study of . . ." or "Observations on . . ."; simply state *what* it is that you are studying.

Abstract
Also known as a summary, the abstract appears on a separate page following the title page. It is a brief summary of the purpose of the experiment, as well as the method, results, and conclusion. For a simple experiment, your abstract may only be a few lines (75–100 words); even for a complex one, you should keep it to about 200 words. (Note: some instructors may not require you to write an abstract.)

Introduction
Sometimes called the *Statement of Purpose* or, simply, *Objective*, this introductory section should include a succinct yet complete statement of your purpose or objective. If, as is often the case, the purpose is to test a hypothesis arising from a specific problem, you should state, clearly and explicitly, the nature of both the problem and the hypothesis. Your introduction should include the theory underlying the experiment and, in general terms, any pertinent background data or equations. Although you may refer to papers relevant to the experiment, it's best to avoid quoting extensively.

Experimental design (or methodology)
This section often is divided into two parts: *Materials* and *Method*. However, some departments prefer that students, especially in first- and second-year

courses, omit the subheadings. Certainly a single section is adequate when the description of materials used is short. The following pointers apply in either case.

MATERIALS

First, list the materials and equipment used and explain how the experiment is set up. If different arrangements of equipment are required, give a full list of the equipment in this section, and then in the *Method* subsection describe each separate arrangement before you describe the respective tests. A simple diagram will help the reader visualize the arrangement of equipment. If the diagram is too large for a regular page, you can draw it on an appropriately numbered and labelled attachment at the back of the report, as an attachment (or appendix), and refer to it within the body of the report. Some departments require that you specify the source or supplier of any reagents you have used. Such information should be included in this section: for example, "Spectroscopic grade carbon trichloride (99 per cent pure) was obtained from B.D.H. Chemicals."

METHOD

Describe the procedures step-by-step, in the order in which you actually performed them. You should include enough detail that others will have no difficulty in repeating the experiment—and, you hope, achieving the same results. (Of course, if you are following instructions in a lab manual, you may not need to copy them out word for word; simply refer to the instructions and give details of any deviation.) When a certain procedure is long, complicated, or not necessary to a full understanding of the experiment, you may describe it in a labelled attachment as an appendix.

Although your description of the experimental method should be concise, it must include all the essential details, so that readers will know exactly what materials to use, when to use them, and what controls to apply if they try to perform the experiment themselves. If you heated a test tube, for example, report at what temperature it was heated and for how long; if you performed a chromatography or other process at a faster or slower rate than usual, state the rate; or, if you used a piece of equipment that had to be set to particular specifications, identify the calibration. You should be consistent also in the way you identify your material (degree of detail, units of measurement, and so on). Identify what statistical tests you have used to analyze your data.

If you have undertaken a number of tests, statistical or otherwise, begin with a short summary statement listing and numbering the tests so that the reader will be prepared for the series. To avoid confusion, describe the tests in the same order in which you have listed them, giving them the same numbers and subheadings.

Experiments always are described in the past tense. When you want to give instructions rather than a description of your method, use the imperative: e.g., "Cut off the top 20 cm of the soil sample."

Scientists writing for scientific journals are divided on whether you should use the active or passive voice. Most prefer the passive voice ("The water flow was started," rather than "I started the water flow") because they believe that its impersonal quality helps to maintain the detached, clinical tone appropriate to a scientific report. Nevertheless, some scientists use—and accept in students' work—the active voice ("I started the water flow") because it is clearer and less likely to lead to awkward, convoluted sentences. Ask your instructor about the department's preferences. Whichever voice you use, remember to strive for *clarity*, *objectivity*, and *consistency*.

Results

This is where you report your observations, data, and calculations. Find out from your instructor whether you are expected to give the details of your calculations or only the results they produced (calculations are sometimes placed in an appendix). In either case, you should pay special attention to the units of any quantities: to omit or misuse them is a serious scientific mistake. Taking care to include all units will also reveal mistakes in your calculations that might otherwise go undetected. If your units don't cancel properly to yield the result you expected, you will know that you have made an error.

Where possible, you should also make sure that the calculated values you report include the "uncertainty" in each of them. For example, you might report that the calculated volume of a hollow sphere is 23.45 ± 0.05 cc (where ± 0.05 is the uncertainty in the volume measurement). Whenever you report any calculations or measurements, check to see whether you need to include the standard deviation, the standard error of the mean, or the coefficient of variation.

The format of the *Results* section depends on the type of experiment performed. Whenever possible, use a computer-based spreadsheet with a quick and efficient graphing capability to summarize your findings. A graph (or chart) usually is preferable to a table because it will have more visual impact, but if you have made several measurements it probably is best to report them in tabular form. When devising a graph without the aid of a computer, be sure to remember the following steps.

- Use properly ruled graph paper.
- Use a scale that will allow you to distribute your data points as widely as possible on the page.
- Put the independent variable (the one you have manipulated) on the horizontal axis, and the dependent variable (the one you measure) on the vertical axis.
- Make the vertical axis about three-quarters the length of the horizontal axis.

And, whether or not you use a computer, remember these three important rules:

- Put error bars (±) on data points, where possible.
- Label the axes, including the units used, so that the reader knows exactly what you have plotted on the graph.
- Title the graph and label it (for example, "Figure 1"), so that you can refer to it by number in your report.

Discussion (or analysis)

This is the part of the lab report that allows you the greatest freedom, since your purpose is to analyze and interpret the test results and to comment on their significance. You want to show how the test produced its outcome, whether expected or unexpected, and to discuss those elements that influenced the results. In determining what details to include, you might try to answer the following questions:

- Do the results reflect the objective of the experiment?
- Did the experimental method help you to accomplish that objective?
- Do the results you obtained agree with previous results as reported in the literature on the subject? If not, how can you account for the discrepancy between accepted values and those you have found experimentally? What (if anything) went wrong during your experiment, and why? What was the source of any error?
- Could the results have another explanation?

For a good discussion, remember to think critically not only about your own work, but about how it relates to previous work.

Conclusion

This is a simple statement of the experiment's conclusion. You don't necessarily need a separate section; the conclusion can also appear as a short summary at the end of the *Discussion* section. Although you may include a summary table or other graphic presentation if it will clarify the conclusion, you should not introduce new information—i.e., data that are not identified within the body of the report.

Attachments (or appendices)

Numbered and titled attachments should include only detailed material that is too extensive to place within the body of the report, such as the diagram and detailed material noted above in the *Experimental Design* section, or extensive statistical analyses.

References

For most reports on work done in the lab, in one day's session, you may not have to include any references. Your instructor will be impressed, though, if you take the time to find and note one or two relevant references (including, perhaps, appropriate sections of the textbook). If the lab test takes several days to run, then you will have time to do a little extra research. You should list the sources of any information used in your report. If you use endnotes to cite some references or present a brief explanatory discussion, you will need to have a separate endnote section, preceding the list of references. If you need to acknowledge a number of sources, you may list them in separate categories, such as *Books* and *Journals*. Be sure to use the referencing style pertinent to your course of study (see Chapter 11). If in doubt, ask your instructor.

REPORT-WRITING STYLE

Scientific reports, like essays, must be written with the reader in mind. Since your reader will be your instructor or TA, you don't need to define elementary terms or explain a method that would be taken for granted by anyone with scientific training. At the same time, you should avoid filling your report with technical jargon when non-technical language will suit just as well. The watchwords are clarity and precision; you want to make it easy for the reader to understand exactly what you mean.

Although the basic rules are the same as for any other kind of writing, scientific reports do pose special problems. Here are a few words of advice.

Avoid using too many nouns as adjectives

Clusters of nouns used as adjectives can create cumbersome phrases:

orig.   adult male kidney disease

rev.   kidney disease in adult males

orig.   apparatus construction

rev.   construction of apparatus

Of course, some nouns are frequently—and acceptably—used as adjectives: *rock* glacier, *reaction* time, *block* stream, *sheet* wash, *stream* channel, *ice* sheet, *loess* plain. Your ear is probably the best judge of what is awkward and what is not.

Avoid using too many abstract nouns

Whenever possible, choose a verb rather than an abstract noun:

orig.   The addition of acid and subsequent agitation of the solution resulted in the formation of crystals.

rev.   When acid was added and the solution shaken, crystals formed.

rev.   When I added acid to the solution and shook it, crystals formed.

Avoid vague qualifiers

As a scientist you must be exact. In particular, you should avoid words such as *quite*, *very*, *fairly*, *some*, or *many* when you can use a more exact term. *Relatively* is especially dangerous unless you are actually relating two or more things.

Avoid unnecessary passive constructions

Although you may prefer to use passive verbs for describing methods and results, you should try to use active verbs in your *Introduction*, *Discussion*, and *Conclusion* sections. Your sentences will be clearer and more direct:

orig.   pH4 is needed for the enzyme.

rev.   The enzyme needs pH4.

> **orig.** It was <u>reported</u> by K.Q. Truong...
>
> **rev.** K.Q. Truong <u>reported</u> . . .

Avoid ambiguous pronouns

The pronoun *it* will cause confusion if the reader can't tell which noun it refers to:

> To germinate, the seed requires water. <u>It</u> must be warm.

Is it the seed or the water that must be warm? If there is any chance of ambiguity, you should repeat the pertinent noun:

> The <u>water</u> must be warm. (or) The <u>seed</u> must be warm.

Be especially careful that the pronoun *this* clearly refers to a specific noun:

> **orig.** When removed from the water, the stalk lost its leaves. <u>This</u> occurred over eight hours.
>
> **rev.** When removed from the water, the stalk lost its leaves. <u>This loss</u> occurred over eight hours.

CHAPTER 7

Presentations and Group Work

This chapter is intended to help you take an active part in seminars, discussion groups, tutorials, and—for upper-level students—conferences. The chapter also offers tips on managing group work, giving formal presentations, and presents suggestions about visual aids, including posters, that you can use in those situations.

PARTICIPATION

The goals of all discussion groups, seminars, and tutorials are insight and understanding, and the key to achieving them is participation. In turn, the key to successful participation is preparation; you don't want to be left fumbling if you are called on to speak. The keys to preparation are reading, thinking, and writing notes. The following pointers will help you to become a good participant.

Preparing

1. **Know what the assignment is!** In most cases, your assignment (say, to read and comment on a chapter) will be announced in the previous week's meeting. If you miss a session, check with a classmate to find out the topic for the next week; remember that your instructor will determine quickly who is prepared and who is not.

2. **Be prepared to be the first speaker.** Even if someone else has been assigned to speak first, remember that plans often go awry. The first speaker will be expected to identify the author's essential argument and to raise leading (most important) questions.

3. **Be prepared to discuss the author's work from a critical standpoint.** Know what you agree or disagree with, or are puzzled by. Never be afraid

to admit that you don't understand something; others may be in the same situation. Insight and understanding will develop through discussion.

4. **Be aware of time limitations.** If you are assigned to be the main spokesperson, your preparation will be critical. If the meeting is fifty minutes long, you may be expected to speak for five to ten minutes—to present the main ideas derived from the assigned reading and to identify some of the questions that came about after considering what the author(s) had to say. Don't plan to say too much, and don't go into too much detail. As the one assigned to get the discussion started, you should identify the high points, raise some questions, and hope that others will jump in with their thoughts and questions. On the other hand, you may have only two or three minutes to summarize your initial reaction to the assignment. In short, it's to your advantage to know in advance how much time is allotted for you to speak. If you do know, practise speaking for that long, either to yourself or—better—to someone else who can be both a time-keeper and a critic. In any case, whether you have a little or a lot of time to speak, it will never be enough to present everything the author discusses; if there were time for that, the instructor would simply have you read the work aloud. But that isn't the purpose of discussion groups. What you need to do is summarize your interpretation of the author's work, concentrating on the central ideas. How can you prepare for that?

The topic

Most assignments for discussion groups, seminars, and tutorials relate to a reading assignment, idea, or quotation that you are to think about and be ready to discuss. You can help the thinking process along by systematically writing down your thoughts and then reorganizing them into a coherent whole. In itself, the act of writing will help you to think.

A READING

Your assignment may be on one or more articles in a journal or chapters from a book. First, scan the material quickly to get a feel for the general points made, the types of research data and methods of analysis used, and the conclusions drawn. If the material is from a journal, there may be an abstract. If so, read it, but be aware that the author's abstract may not convey the information you need to summarize the work.

The next step is to read the whole assignment with care, jotting down the major points as you go. It's essential to grasp the main lines of the argument before you begin reflecting on what you have read. Although some questions may be answered by the author, others may not, so be prepared to pose those questions in the group meeting.

You'll find that you can participate much more freely if you have prepared some short notes to use as memory-joggers. But keep them brief: lengthy notes will tempt you simply to read out what you have written. Sometimes it's useful to write down a short quotation too, if you think it may be appropriate to cite in the discussion.

AN IDEA

If you are asked to discuss an idea rather than a specific piece of writing, it will probably be presented in the form of a quotation or a brief statement. Such assignments can be more difficult to prepare, but there are some steps you can take that should help:

1. **Think about the context.** In the case of a quotation, try to find out more about the source from which it was taken. If the source is a book or journal in the library's collection, examine it quickly to determine the context of the quotation. Even if you decide that the quotation can be considered without reference to its source, you will certainly have to consider the context of the course in question (see point 3 below).

2. **Think about the contents.** Is there one sentence or several? Is there one idea, or two, or more? If there are several sentences and ideas, identify the main point first; then look at the remaining points in relation to it. What is the author saying? Is the point clear, or is there a subtheme that is not obvious at first glance? Does the author seem to be working from an assumption that you find questionable? Write out your understanding of the author's main point.

3. **Think about the author's statement in relation to the course.** Is there an obvious connection to the readings assigned to support that week's lecture(s), earlier lecture material, or class discussion? There may be a link to material examined several weeks earlier. Think about the sweep of material covered in the course to date. Is there a summarizing quality to the quotation? Does it grasp some essential point that the instructor may be aiming at?

4. **Try writing a one-page summation of your understanding of the quotation and your responses to it.** In contrast to an abstract, which demands objectivity, this exercise includes a subjective element: not only your understanding of the quotation itself, but your exploration of your response. It is a

combination of statement of fact and personal reflection. From this exercise will come more questions, which in turn will trigger further thought.

In class

Your class likely will be meeting in a room that allows the participants to sit in a circle, perhaps around a table. Take a pad of paper with you so you can jot down the points made by others in the group—their conclusions and their questions. Try not to become totally immersed in note-taking, though: you want to keep your head raised as much as possible, for two reasons. First, it will help you to see people's faces and/or hear their inflections, both of which can be crucial in conveying meaning. Second, you'll find it easier to participate if you are focusing on the others in the group rather than your note pad. Looking at people as you talk allows you to see if they are listening to you. If you notice that someone is "turned off," you might try asking that person a question. Above all, speak clearly so that the other participants can hear what you have to say.

WORKING IN GROUPS

Being able to work effectively in groups to produce a research report or other group assignment is an important ability that your instructors (and future employers) value highly. Successfully completing a group project is partly a result of organizing and managing all the tasks required, so the material that follows provides suggestions about how to plan your group projects using a basic Gantt chart.

All group projects involve a set of activities that include both sequential and parallel tasks, have distinct start and finish dates, and involve constraints on time, people, and possibly equipment. A Gantt chart is one useful planning tool that may be used to help you work out how long it will take to complete your project on time and the sequence of tasks that your group must carry out to achieve that goal. When a project is underway, Gantt charts help you to monitor your group's progress and to identify any remedial actions that may be necessary to bring the project back on schedule. Developed by mechanical engineer Henry Gantt in the 1910s, the Gantt chart may be prepared manually or by using computer software packages. The chart displays a time scale on the horizontal axis and the rows of bars on the vertical axis show the beginning and ending dates of individual tasks in the project.

The basic form of the Gantt chart is shown below: each task in this three-week project began when the task above it was completed.

Task	Duration	Week 1	Week 2	Week 3
1	1 week	▬▬▬▬▬		
2	1 week		▬▬▬▬▬	
3	1 week			▬▬▬▬▬

Figure 7.1   Basic form of a Gantt chart

To plan a group project using a Gantt chart, you can follow these steps:

1. Identify **and list all the significant project activities that need to be completed** (e.g., identifying the field site, searching the literature, etc.). For each task, show the earliest start date, how long you estimate it will take to complete, the team member(s) who will be responsible for each task, and any resources required. Note that the timing of some of your tasks may be dependent on other activities being completed first (sequential tasks), while other (parallel) tasks may be done at any time. Initially, it can be difficult to estimate accurately the durations of various tasks; often, doubling your guess works best! If you are preparing your Gantt chart manually, use a piece of graph paper to show your estimated time schedule (days or weeks) through to completion of the project. Table 7.1 shows what a work breakdown could look like (these same steps are used in Figure 7.2):

2. **Plot each of the tasks onto the graph paper**. Starting at the earliest possible date, draw each task as a bar, where the length of the bar represents the estimated duration of each task. The left end of the bar represents the expected start date of your project and the right end marks the expected completion date. Above the bars, identify the time taken to complete each task (e.g., specify the number of days and place in parentheses). Specify a reasonable number of tasks (15 to 20 maximum) so you can fit your chart onto a single page. If you have a more complex project, you can split it into main and subtasks, and complete an overall Gantt chart for the main tasks and separate charts for the subtasks. Don't worry about scheduling each task yet, and don't be surprised if the first draft of your diagram looks untidy.

3. **Schedule the project activities**. Take the Gantt chart and schedule the project tasks so that the sequential activities are carried out in the required sequence, ensuring that the parallel tasks do not start until the activities

Table 7.1   Work breakdown (selected elements only)

Task description	Possible start	Time estimate[1]	Who is responsible	Type/ dependent on[2]	Resources
1. discuss project and its objectives	week 1	2 days	MN, DK, EJ, AS	Sequential/—	
2. site selection; field verification	week 1	2 weeks	DK, EJ	Sequential/ task 1	
3. identify equipment needs; obtain equipment	week 2	1 week	DK, EJ	Parallel/ tasks 1, 2	stream gauges, GPS . . .
4. conduct literature review	week 2	7 weeks	MN, DK, EJ, AS	Parallel/tasks 1, 2, 3, 6	
5. collect data	week 4	2 weeks	DK, EJ	Sequential/ tasks 1, 2, 3	
6. progress report due	week 5	1 week	MN, AS	Parallel/tasks 2-5	DVD projector
7. analyze data	week 6	3 weeks	MN, AS	Sequential/ task 5	statistical software package
8. write final report	week 8	4 weeks	MN, DK, EJ, AS	Parallel/1-7	

Notes:

1. Time estimates are illustrative only and are not recommended timeframes.

2. Whether tasks are sequential or parallel largely depends on context.

on which they depend have been completed. The timeframes for some tasks may overlap. Try to make the best use of the time each group member has available, remembering that, ideally, each person should contribute an equal amount of time to the project, though not necessarily to each task. Ensure that you identify important milestones within the chart, such as when a progress report or presentation is due, and the final due date of the group project. Milestones or checkpoints may be marked by a special symbol, often an upside-down triangle.

It's advisable to include a little "slack time" in the schedule to allow for delays, including equipment and computer problems or even the failure of one or more group members to complete their tasks precisely on time. If a group member appears not to be putting an appropriate level of effort into the project, and if your efforts to contact or speak with the person are unsuccessful, consider talking with your instructor as soon as possible; he or she may take additional steps to help ensure that member's participation in the project. If the problem is not resolved, and if your instructor provides a peer evaluation opportunity at the end of the project, having spoken about your concern previously will help the instructor to assess more accurately every member's contribution to the project.

4. **Monitor your progress**. As the project progresses, regularly update your chart by filling in the bars to a length proportional to how much work has been accomplished on each of the tasks. You can get a "quick read" of overall progress on your project if you draw a vertical line through the chart at the current date. The filled-in portion of the bar representing tasks that have been completed will lie to the left of the vertical line, as will the filled-in portion of the bar for tasks that are behind schedule. Current tasks will cross the line, and ahead-of-schedule tasks will have the filled-in portion of the bar to the right of the vertical line. Future tasks are positioned completely to the right of the line.

A simple example of a completed Gantt chart might look like Figure 7.2

PRESENTATIONS

Many courses require that each student give a formal presentation on some research conducted during the term. Such presentations are often initial explorations that can be reworked before they are written in final form for the instructor. In fact, you may find that presenting your material in public triggers thoughts quite different from those generated by writing alone. Consider the following: there is the formal presentation you plan to give, the one you actually give, and the one your audience hears! You need to be sure that what you actually give conveys what you really want your audience to know. Careful preparation is the key—and good delivery, of course.

Specify project title here

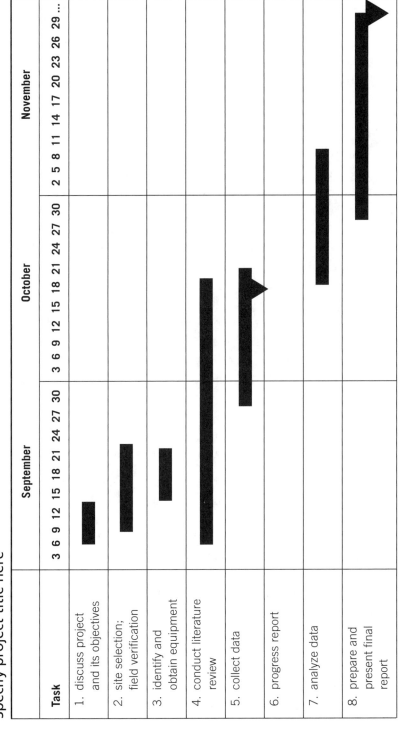

Figure 7.2 A simple example of a Gantt chart

Oral presentations can be daunting if you have no previous speaking experience. Keep in mind, first, that you are among friends, all of whom will have to make their own presentations; second, that experience in giving presentations can be invaluable in later life. Here are several pointers that will help you to structure your talk.

Preparing

1. **Know your topic.** An effective talk is always based on solid research. For your presentation to be believable, you must become "the expert." So, within the time constraints, learn as much about your subject as possible. Having a good grasp of the topic will boost your self-confidence.

2. **Be aware of what your audience needs to know in order to understand what you are going to say.** Does your audience already know the definitions, concepts, and processes that you will be referring to? If not, identify them clearly. Start by giving the title for the talk; then pose a question, or at least define your aim. It may be appropriate to remind your audience that your topic relates to some issue, concept, or broader area of discussion that has previously been examined in class. If you are reporting on field work, be sure to identify the location and character of the area where the research was done.

3. **Do not write out everything you plan to say.** There are several things you can do to help yourself as you speak. Consider the following tips:

   - Use index cards or paper. Number them, in case they are dropped and you must get them back in order quickly.
   - Using headings to group your material, list the main points, supporting details or evidence, quotations, and any questions you wish to pose. Include all the main points you wish to make, not just some of them.
   - If possible, use visuals to convey any detailed information rather than trying to read it out.
   - Number your points. This will help you avoid anxiety over what comes next.
   - Underline key words or points that you want to emphasize.
   - If you wish to make a side-comment or joke (*if* appropriate), jot it down too, in case your memory blanks out at the key moment.

- Keep your notes simple. Don't have long run-on sentences. Use point form.
- Use large print (or a large font size) so you can see the material easily if you have to move away from your notes to put something on the blackboard or if you have to move to the overhead projector.
- Don't put too much on one page or card. Leave a large space at the bottom. This will help you avoid looking down as you approach the foot of the page or card.
- Try not to be glued to the written material. Use it as a guide to your spoken word. Reading directly all the time tends to produce a monotone presentation, and hampers communication with your audience.
- Be aware that if you provide a handout or use a visual on a screen, it may distract your audience. Be sure that whatever you give them in this manner is totally relevant to your presentation and that it can be read quickly. Refer to the material as you speak, thus guiding your audience to what you want them to get from it. In short, think carefully about what to include in any such material.

4. **Outlines for others may be required.** You may be asked to provide a handout for each student: sometimes a bibliography, sometimes an outline of your presentation. If the latter, keep it shorter than your speaking notes. Such outlines should be in an orderly form and must include your name, the course number or title, and the title of your presentation. As you speak, you can glance at the outline from time to time as a constant reminder of where you are in your presentation, even as you also refer to your more detailed notes. Even if you aren't asked to do so, it's a good idea to give other students a list of some important references that pertain to your topic; not only will it give them some useful information, but it will help to remind them of your contribution to the class.

5. **Use visual aids when possible.** Visual aids can enliven and inform a class, but they can also be distracting. For topics on *posters*, see pp. 116–20; *computer-generated images*, see pp. 111–12; and for *maps*, *tables*, *diagrams*, and *graphs*, see Chapter 8. Before proceeding, here are four general points:

- When possible, use a horizontal image. If you must use vertical visuals, make sure the screen is high enough for people at the back of the room. If the screen is too low, they won't be able to see the bottom.

- Keep all visuals simple, clear, and focused on a point; multi-purpose visuals may only confuse your audience.
- Make sure the text you use in any visual (including maps and graphs) is large enough to be seen from the back of the room and, as with slides, be sure that the image is in focus.
- If you have a lot of numbers to show, consider plotting them on a graph that will be easy for your audience to grasp.

There are several types of visual aids. Here are a few comments about each of them:

- **Blackboard.** If you plan to use a black—or green or white—board, you may need to put some of the material on it before the class meeting: you don't want to spend too much time with your back to the audience. If you intend to draw suggestions from others in the class, writing their comments in summary (or point form) on the board can be a useful trigger to later discussions, but you may want to have a designated assistant do the actual writing for you.
- **Overhead projections.** Overhead projections of an outline of your talk, or quotations, or maps and diagrams can be helpful. Your illustrative material can be copied onto transparency film using a photocopier—or a computer and a high-grade colour printer—and shown in class with an overhead projector. Overheads have several advantages: they are often clearer than work done on a blackboard, especially in big rooms; they can be prepared in advance and still written on (using special pens) during the talk; they can be partially covered and moved on the screen as need be; and, since they are ready to use, the images on the screen can be changed very quickly. Before the presentation, stack your images in the order you plan to show them and know which side to place on the machine—you will lose your audience's attention if you have to hunt for the right image or if you place one upside down. Since you can write on the plastic used for overheads, they may be useful for summarizing points made by you or others. (They are especially useful if you are allergic to chalk!)
- **Slides.** Slides of scenes, people, or equipment may be useful too— but only if they actually show clearly what you think they do! All too often, people who take photos see something in them that others do not. If you are unsure about the effectiveness of your slides, show them to colleagues before the class presentation and have them

tell you what they see; their answers may not be what you expect. In that case, you may want to use different slides, if any at all. If you are using a 35mm projector, be sure to load the slides into the carousel (the circular rotating holder that loads photographic slides into the projector one at a time) correctly, paying attention to the way each slide faces (to avoid it being back-to-front) and orientation (so it is not upside down). *Before* class, check the slides (preferably using a projector) so that during your talk you are not caught off guard by a slide that is oriented incorrectly or is out of order. If you intend to use slides stored on a computer disk, check to see that they are in the order you wish to project them (using the projection equipment available to you in the classroom).

- **Movies.** You may be able to show a movie to a discussion group or class. Only select something that is totally pertinent to the situation. A short section of a movie may be all you need show in order to make your point. Check with the appropriate library staff about any copyright limitations, which might preclude you from using a movie in a classroom setting.

6. **Computer-generated images.** If you have access to the necessary equipment, computer-generated images (using a program such as PowerPoint) can be projected directly onto a screen. Some lecture and seminar rooms are equipped for this; you just hook up your laptop. If this is not the case, check with the department's chief technician to see if "in-house" aid is available. Otherwise, your college or university likely has a teaching support/communications media service that can provide equipment and assistance.

Computer-generated images can be combinations of slides of places, text, graphs, and maps. Since they can be stored easily on your laptop, or on a CD, DVD, or memory stick, they are readily transportable. If you are not experienced with the relevant program or equipment, some of your presentation time might end up wasted while you figure out the technology. If a course is available on the techniques, take it, and apply what you learn. With practice, you will find it is possible to generate useful images that can be shown in sequence. It also is possible to jump back to a previously seen image before returning to your ongoing sequence though, for some, this can be both cumbersome and distracting to the audience.

As you prepare your material, consider the following guidelines.

- Be consistent in the styles you use for lettering, formatting, and colours.
- Use a plain background for all of the images. Don't overuse colours (e.g., different colours for each slide). White lettering on blue can be effective. Avoid using yellow and orange for words, because they are hard to read, but they may be used to show a distributional pattern on a map.
- Don't use several colours on an image unless there is a clear and understandable reason to do so.
- Choose a simple, clean font that can be read easily from the back of the room. Sans serif fonts work best for onscreen presentation material such as overheads and PowerPoint slides.
- Use a size of type that can be read with ease from the back of the room. You may need to try several sizes before identifying the best size for your purpose.
- Don't overload any one slide with too many words. Use point form for succinct statements. Avoid embellishments. If anything is not essential to the information to be conveyed, don't include it. Normally, each slide should make one point, or two or three (bulleted) related points.
- If you show a quotation, be sure to include quotation marks and identify the source. Given space limitations, you may list only the author's name, but be prepared to give the full citation after your presentation, if called upon to do so.
- Keep graphs, tables, and maps as simple as possible. Don't overload them with too much data.
- Finally, don't be seduced into using only the ready-made templates on PowerPoint. Think differently. Know what *you* want to show, why, and how you want it to appear. Format your own slides—then you can be sure that you will have what you want.

7. **Practise giving your talk.** Know the time limit, time yourself, and test your notes. Are they adequate? Is there too much information, or not enough? Practise using your visuals as you speak so that you know when and how to use them and how long you can spend on each. Keep the specific purpose of each item in mind and be aware of the next point you wish to make. Whatever aids you use, remember that if you take longer than planned on one item, you will have to rush later ones. Be aware, too, that

you will probably cover your material (and show visuals) faster in a "dry run" than in class. If possible, practise with a friend who can give you feedback. Above all, the keys to preparing for any presentation are good thinking, good writing, and good preparation.

Presenting

1. **Dress comfortably (but not sloppily).** Be wary of trying something new for the occasion: the day you give a presentation may not be the best time to wear a tie if you aren't accustomed to one, or to try out new shoes that don't quite fit.

2. **Be on time.** If you don't arrive until the last minute, you may find that another student is absent and the schedule has been changed to slot you first. Also, if you plan on using any equipment, you'll have to set it up and make sure it is functioning properly before the session begins.

3. **Start by identifying what you are going to do.** Tell your audience what your topic is and briefly outline how you plan to present it. If your time limit is very short, don't give an outline—just present the essential information.

4. **Keep track of time.** Note the time on your watch (place it on the table where you can see it easily) when you start your presentation and periodically check it to see how much longer you have. An analogue watch, which shows relative time, is more useful than a digital watch.

5. **Speak loudly, clearly, and with enthusiasm.** Everyone in the room needs to hear you. Pronounce your words distinctly (getting rid of your chewing gum will help!). Try to speak with enthusiasm; a monotone can be deadly. Pace yourself. Don't speak too quickly.

6. **Never apologize.** Comments such as "I'm not really ready to give this talk," "I don't feel well today," or "I'm not very good with this machine," will irritate any audience. They are there to learn about your topic, not the state of your health or preparations.

7. **Look at your audience.** Eye contact will help you see whether your audience understands what you're saying. Puzzled looks may be a sign that you need to repeat or rephrase a point to make it intelligible.

8. **Use your visual materials.** Having thought carefully in advance about how you intend to use your visuals, proceed in a way that will keep your audience focused on your essential message, not just on the supporting material. Consider the following tips, which assume you are using computer-generated slides (though the points also pertain to the use of overhead and 35mm slide projectors):

- Don't lose precious time turning on the machine for the first slide during your presentation. Have everything switched on and ready to go before you start.
- Start with a title slide that contains the title of the presentation, your name, and, if you think it is appropriate, the name of the course and the date.
- Your second slide should give a short outline of your talk so the audience knows what to expect.
- Don't overwhelm your audience with too many slides. A good rule of thumb is one slide per 40–60 seconds of your talk.
- Avoid the temptation to speak too long about any one item.
- Don't obscure anyone's view by standing in front of the screen, and try not to turn your back on the audience. Presenting from either side of the screen usually works bests.
- If you choose to show a slide with several small images, or many numbers, be sure to leave it on the screen long enough for the viewers to grasp them all; otherwise you will lose their attention, or at least leave them frustrated.
- End with a point-form summary, or conclusions, slide.
- Be wary when using some of the "sliding" and "folding" change methods that are possible as you progress through your slides. The process of changing the images may become more interesting for the viewer than the content of the images or, for others, the process may simply be irritating. Either way, you could lose the attention of your audience. The key? Keep the process simple.

9. **Use a numbered outline as a memory aid.** Even if you don't identify the numbers to your listeners, numbering the points on your outline will make it easier to find your place when you glance at your notes after looking up and around the room.

10. **Use humour.** A joke can help to reinforce a point, if it is relevant. Just don't turn your presentation into a stand-up routine—and never, ever tell a joke that includes racial or other slurs.

11. **Pose questions for your listeners as you speak.** Questions will help to keep them alert.

12. **Summarize your main points and then draw the presentation to a conclusion.** Your conclusion should bring the audience back to the main issue or question(s) raised during your talk. Try to end with a question (or a controversial observation) that will stimulate your audience to respond.

13. **Finish on time.** You don't want to be told by your instructor that "time is up" when you still have information to share. Remember to place your watch on the table so you can see it. Keep within the specified limits and don't expect to be granted additional time to finish.

14. **Be prepared for questions.** It's a good idea to repeat any question so that everyone can hear it before you respond. Keep your answer short and to the point. If you didn't hear or didn't understand a question, don't be afraid to request that the person repeat or clarify it. If you can't answer a question, you can always say you don't know and ask whether others in the group have any ideas; or you can refer the question to your instructor.

Tutorials

Although most of the above points also apply in tutorials, there is one fundamental difference between these small groups and large seminar or discussion groups. Whereas the latter usually have anywhere from twelve to twenty participants, a tutorial normally will involve only the instructor, you, and perhaps one or two other students. The intimacy of such a small group means you cannot escape behind someone else—you simply must be prepared. The structure of tutorials varies from instructor to instructor, but you always will have to read carefully and be prepared to discuss your reading. Make sure you keep your reading notes throughout the term (with a full record of author, title, and source information), since you may have to write a summing-up paper, usually in the form of a literary review, as part of the tutorial requirement.

GIVING A PAPER AT A CONFERENCE

Upper-level undergraduates and, more often, graduate students may be invited to give papers at conferences, sometimes with a professor as co-author. Many of the comments in this chapter apply to conference presentations as well as class work. Rehearse thoroughly, speak in a structured but free-style manner (don't read your paper word for word), enunciate clearly, speak directly to the audience, keep within the allotted time—and be ready for questions!

Posters

A poster is a self-contained visual display that generally uses a mix of narrative and visuals (e.g., maps, tables, graphs, and pictures) to convey information. In some courses posters may be required to display findings from a field project or as part of a senior undergraduate project, but they can be useful in other situations too, including labs, seminars, and departmental "open houses." In addition, posters are used frequently at professional meetings, where instead of giving papers, participants may stand beside their posters and answer questions from other conference participants.

The dimensions of the poster may be determined for you, either by your instructor or by the size of available display boards. Some instructors will have you use the dimensions required by learned societies, e.g., for the AAG, 1.2 metres by 2.4 metres (about 4 feet by 8 feet). Creating an effective poster—one that is unified and coherent—requires careful planning. Here are some pointers.

BASIC ELEMENTS TO INCLUDE
- Subject title
- Author(s) and, if needed, affiliation
- Textual material: statement of problem, research methods and data used, evidence and results, conclusion
- Visual material: maps, graphs, tables, photos, as required
- Short list of pertinent references
- Acknowledgements, if appropriate

"VIEWABILITY" AND "CATCHABILITY"
- The title of your poster should be in print large enough to catch someone's eye from at least 4 metres (about 13 feet) away.

- Other material should be legible from a distance of about 1.2 to 1.5 metres (4 to 5 feet); 18-point type is the absolute minimum size. Also, remember that text printed all in upper-case letters is hard to read.
- Related to "viewability" is "catchability": your poster needs to grab the attention of potential readers. Good overall design is essential. Vibrant colours can be effective in attracting attention, but to keep the reader's attention they should be used selectively.

TEXTUAL MATERIAL

- Be concise. Editing—and re-editing—are required to make a point with brevity and clarity.
- If possible, use point form rather than paragraphs of text. Constantly ask yourself whether the textual material you want to present is essential; if the answer is no, delete it. Your final word count should be between 250 and 450, with the lower limit likely if there are several visuals.
- If you have information that you believe is essential to your case but is too detailed to include on the poster, prepare a handout for people who want to know more. Bear in mind, though, that if the assignment was to prepare a poster, with no addenda, your instructor may not consider this "extra" material when grading your work.
- Experience has shown that readers—other than your professor, of course—will tend to start with the title, glance at your statement of problem, and then read your conclusions. Only if you have captured their interest will they consider the rest of your material. Keep this tendency in mind as you decide on the poster's title and write your conclusion.

VISUAL MATERIAL

- Include only those visuals that directly inform readers about your project. Don't include tangential illustrations just because they are "nice."
- Keep your visuals as "clean" and simple as possible, so that readers can grasp their essence quickly.
- If you must use multi-purpose visuals, ensure that your graphic selections are clear and distinct, so that readers can understand the essential information easily.

- Consider presenting your data on a graph or map rather than in a table.
- Ask yourself whether the visuals convey essential information. Have you chosen the best type of graph? Are the patterns on maps and graphs clear and readily understood? Presenting different sorts of information in different forms will make it clear that they are discrete. If in doubt, make separate visuals.

LAYOUT

- Use graph paper (cut proportionally to the dimensions of the poster board) and small pieces of paper (representing your text and visuals) to plan the layout of your poster.
- Identify everything that you wish to include (text and visuals). Roughly identify the size of the space they will occupy and cut small pieces of paper accordingly. Placing these on graph paper will indicate whether or not you have been precise enough.
- Lay the pieces of paper on the graph paper in various arrangements to see what works best. Keep the figures, tables, and text blocks balanced (see Figure 7.3).
- Place your explanatory text blocks close to the pertinent visuals.
- To help your readers move easily from one section to the next in a logical manner, use columns and rows, with sequential numbers or letters as guides. You also could use arrows or pointing hands.
- Don't overload your poster. If you have too much material, edit again. Aim to make your presentation increasingly more precise, especially the textual material. Edit and rewrite until the message of your text is absolutely clear, using as few words and, consequently, as little space as possible. Accept that posters cannot convey as much information as written essays or oral presentations.
- Leave additional blank space as a frame around sections you want to highlight.
- Finally, before you paste everything together, check to see that your visual materials are legible from at least 1.2 metres (4 feet) away.
- If all your materials (text, photos, graphs, and maps) are in digital form, or can be converted to it, you may be able to assemble your poster on a desktop computer using an illustration program such as Corel Draw®. Access to a large-format printer (either in your department or at a printshop) will permit you to print your work on a single large sheet.

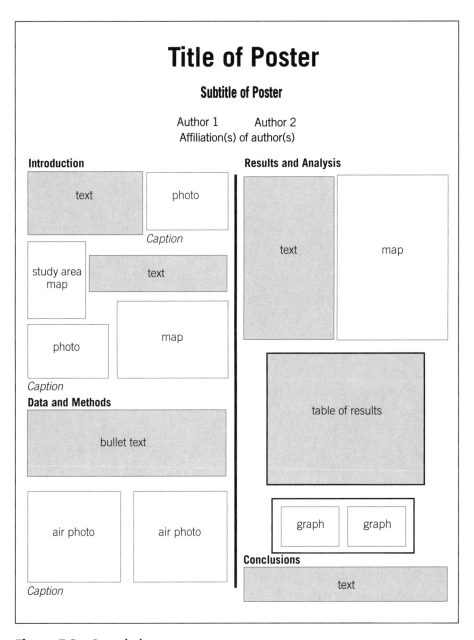

Figure 7.3 Sample layout

CARING FOR YOUR POSTER
- Posters can be awkward to carry, especially if you have to walk any distance through rain or snow. To give your poster a protective cover,

you can laminate it. Check with your department's technician to see whether you can have this done "in-house" (undoubtedly for a fee); if not, you can get it done commercially.

- Go to the session prepared: take your own pins and tape!

CHAPTER 8

Illustrating Your Work

Not all of your course work will require writing. Some kinds of data are most effectively communicated in tables, equations, diagrams, graphs, and maps, and some types of analysis cannot be done without such "illustrations." *Graphicacy* (dealing with graphic images) and *numeracy* (dealing with quantifiable data) are as important as *literacy* (dealing with written language) for students in the environmental sciences generally, and in geography specifically. Visualization and experimentation are important parts of the creative process. Like writing, creating a visual illustration is a complex process that takes time and reflection. A successful illustration will convey something about your data that might not be captured as quickly or effectively by words—if words could convey it at all.[1] Care must be taken, however, to ensure that your illustrations represent the data accurately.

This chapter briefly identifies some basic considerations you will need to keep in mind as you create mapped, tabular, and graphed illustrations. Helpful suggestions are provided on *maps, computer-generated images* (see also pp. 111–12), *GIS-based projects* for publication on the department's website, *tables, equations, diagrams,* and *graphs*. On *slides* and *photos,* see pp. 110–11, 154; *posters,* see pp. 116–20; *sketch maps* and *sketches,* see pp. 152–4.

THE BASICS

Don't be deterred from designing and drafting your own maps and diagrams because they can't meet the standards of the illustrations presented in this chapter, most of which were compiled by professional geographers and produced in final form by professional cartographers. With practice, both your ideas and your execution will become more refined. Careful compilation of data and consideration of what you want to show, along with thoughtful composition and clean, uncluttered final copy, are the keys to success. Two basic approaches are possible:

- For some work—especially in the first year of your studies, or during field exercises when hand-held or laptop computers may not be available—you will be permitted to use traditional techniques and equipment: sharp pencils, clean rulers (with unblemished straight edges), compasses, and graph paper. Pen and ink and a light table will improve the quality of your final drawings.

- For other work—especially at advanced levels of study and for theses—you will be expected to use computer programs to analyze data and create visual displays of the results. You will be able to use some elementary computer programs without instruction (though the maps and graphs produced by these programs are generally of poor quality). Training in the use of more advanced programs will be available through courses offered by your institution's geography department.

For all visual materials, the following points are fundamental:

- Know clearly what topic you want to illustrate, and keep to it.
- Know why you want to include the illustration. What will it contribute to your paper or poster?
- Since poor data will not make useful illustrations, make sure that your sources are reliable.
- Think carefully about how best to show the data: with a graph (which type?), a map (what kind?), a diagram (which style?), a table, or an equation? Can the data be shown on a line graph? Or can they be aggregated into discrete groups to be plotted on a bar graph or listed in a table? Or do they have a spatial aspect that will be most clearly shown on a map?
- Don't try to cram too much information into an illustration; otherwise it may be difficult to follow.
- Make sure your illustration contains all the information necessary for it to be intelligible on its own and yet also relate to the text, by expanding upon or clarifying a point.
- Refer to each illustration by citing its number in the text: maps, graphs, and diagrams are all figures (hence "Figure 1"), whereas tables are listed as such ("Table 1").
- Identify your source(s). If you have taken material from someone else's work, state that it is "from . . ." or "excerpted from. . . ." If you have taken material from one or more sources and manipulated it in some way, state that it is "based on . . ." or "adapted from. . . ." If you are presenting your own original data, gathered from interviews,

field work or laboratory observations, be sure to give yourself credit in a manner acceptable to your instructor or supervisor: for instance, "author's field work" or "original data gathered by the author."

The following sections offer suggestions on several types of illustrations.

MAPS

Maps show spatial patterns and relationships. They can display either data required for the purpose of analysis or information derived from the analysis. Geographic information about the (actual and perceived) contents of the earth's surface is recorded in three basic forms. *Points* represent things that have discrete locations: depending on the scale, examples range from monuments to houses or from schools to towns to cities. *Lines* represent things that have length, such as roads, railways, rivers, and political boundaries. *Areas* represent features such as parks or lakes, or distributions (for instance, of soil types).

All maps are selective. They include only the data that the compiler wants them to show; in other words, they present an abstraction of reality. Since the compiler may (knowingly or unknowingly) misrepresent data through the use of inappropriate projections or scales, poor classification of data, inaccurately manipulated data, inappropriate symbols, poor selection of colour or shading, incorrect plotting of locations, or omission, maps also can lie.[2] Thus, care must be taken to ensure accuracy. As with written work, you need to think about the most effective way to express your data visually, what data to include, and how best to display them.

Location maps generally are quite simple: they show the absolute and relative locations (i.e., sites and situations) of one or more features within a particular locale or region. Other types of maps have other uses. For example, Figure 8.1 shows patterns of land use, identifying the distribution of buildings in an older area of a large city by selected categories (using point symbols) and by heritage value, actual and perceived (using a combination of area shading and point symbols). Figure 8.1 offers good examples of large-scale maps: they cover a small area.

In contrast, Figure 8.2 covers a much larger area, the entire continental U.S.A. Clearly, the degree of generalization is very different. Each computer-generated map shows a distribution. The two maps together reveal time-space changes. Specifically, for the years 1995 and 2004, the maps identify the locations, and thus the distribution, of Wal-Mart superstores (averaging 181,692 sq. ft.) and of the regional distribution centres that fed them. The maps not

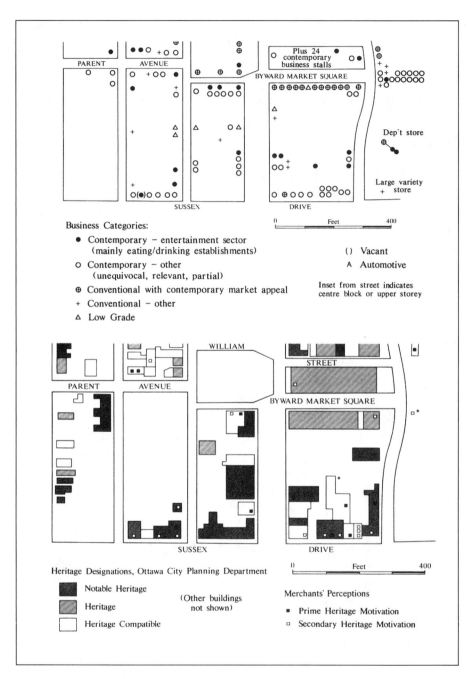

Figure 8.1 Heritage properties in Lower Town, Ottawa

SOURCE: J.E. Tunbridge, *Of Heritage and Many Other Things* (Ottawa: Carleton University, Geography Discussion Papers, No. 5, 1987), pp. 4, 19.

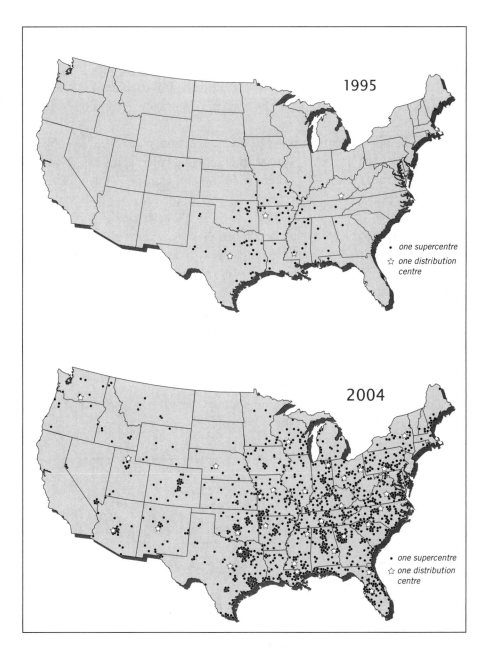

Figure 8.2   **The distribution in the U.S. of Wal-mart superstores and regional distribution centres in 1995 (top) and 2004 (bottom)**

SOURCE: Thomas O. Graff, "Unequal Competition among Chains of Supercenters," *Professional Geographer*, Vol. 58 (Blackwell Publishing, 2006), pp. 59, 60; by permission of the author and publisher.

only reveal the changed distribution but, implicitly, they raise questions about how and why the areal spread occurred as it did between 1995 and 2004. Even a quick glance is enough for the basic patterns to be identified. How many words would be needed to convey the same information that is shown in the two maps? In fact, it would be hard to describe in words alone what is shown on the maps.

Maps can be part of the process of analysis, helping you to solve problems by suggesting insights that otherwise might escape you. Such was the case in some research that required an understanding of the order of amendments to a motion introduced in 1856 in the Parliament of the Province of Canada (for example, Figure 8.3). The order made sense only once the spatial dimension was explored with the help of a map. When the constituencies (the many sub-units within Figure 8.4) of the politicians who proposed the amendments were plotted (using the dark striped pattern) and the order of the original motions and the amendments were identified (using numbers, with arrows to identify the locations being suggested as alternative sites), it became evident that the amendments were put forward mostly by Ottawa-region representatives as part of a scheme—which ultimately succeeded, but only in 1859—to have Ottawa named Canada's capital (although the city didn't become the effective seat of government until 1865–6).

Similarly, mapping data on voting patterns, whether aggregate data from a general election (by voting station, constituency, or province/state) or indi-

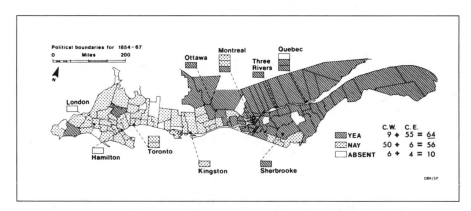

Figure 8.3   A motion to declare Quebec the permanent capital of Canada, Legislative Assembly, Province of Canada, April 16, 1856

SOURCE: D.B. Knight, *Choosing Canada's Capital: Conflict Resolution in a Parliamentary System* (Ottawa: Carleton University Press, 1991), p. 178.

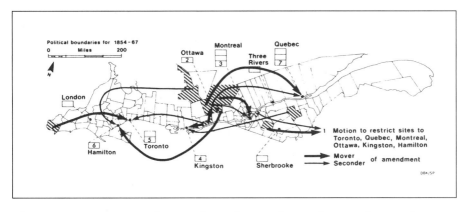

Figure 8.4   Movers and seconders of a motion and several amendments, Legislative Assembly, Province of Canada, April 16, 1856

SOURCE: D.B. Knight, *Choosing Canada's Capital: Conflict Resolution in a Parliamentary System* (Ottawa: Carleton University Press, 1991), p. 160.

vidual votes within a legislature, can reveal patterns that otherwise might not be discernible. For example, Figure 8.3 reveals something of the territorial attachments among those same early parliamentarians. More specifically, the mapped results of votes cast (yea and nay) by each member of the legislative assembly revealed the existence of two major "regions of the mind" within the territory in question, regions that were not readily evident until the data were mapped. Of course, the patterns then had to be interpreted using information from newspapers and other sources, for the votes cast in this case contrasted markedly with the more intricate pattern of individuals' political party affiliations. This was one of more than a hundred such maps used to analyze the legislators' local, regional, and "national" attachments to place during the Canadian debate.

Maps also may show distributional patterns and quantities: for example, a network of settlements showing their sizes and the routes between them, or the varying types and yields of agricultural production in a given region. Figure 8.5 (which includes an inset map showing the relative location of the research study area) identifies n'Daki Menan, the ancestral homeland and geopolitical territory of the Teme-Augama Anishnabai. The named traditional family territories—which would be hard to describe in words alone—are delimited by the broken lines. By contrast, the area and bounds of the Wendaban Stewardship Authority follow the straight-line township boundaries imposed by government. Major bodies of water, highways, and towns are identified also. Note the inclusion of scales and directional indicators.

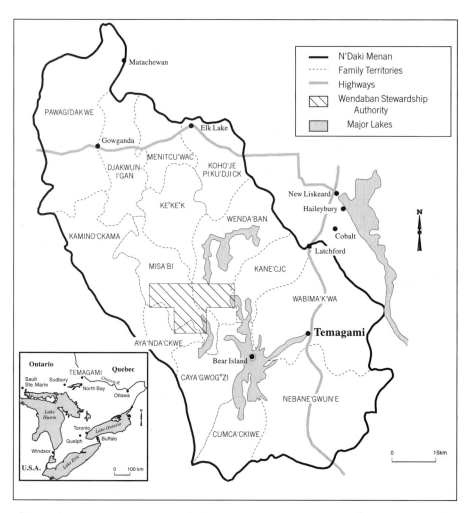

Figure 8.5   The traditional lands of the Teme-Augama Anishnabai and the areal extent of the Wendaban Stewardship Authority

SOURCE: Recompiled from Jeremy Shute and David B. Knight, "Obtaining and Understanding Environmental Knowledge," *Canadian Geographer,* Vol. 39 (1995), pp. 101–11.

A standard method for compiling and drafting maps

If you are not yet able to produce computer-generated maps, you can still fulfill course requirements using drafting paper, graph paper, mylar, a sharp pencil, Letraset, a ruler, squares, and a compass. Before proceeding, however, you should take into account the following standard methodological points, which apply to all maps:

- **Delimitation of scope.** What is the purpose of the map? Will it explain or expand on a discussion in the paper? What exactly do you wish to show? What will you *not* show?
- **Scale, projection, and orientation.** What is the most appropriate scale for the particular topic and data, so that you can include all the detail required? For a small-scale (large area) map, what is the most useful orientation (see p. 39)? (Since north is usually at the top, you will have to justify any change from this pattern).
- **Selection of items to be included.** Is all your material pertinent, or only some of it? Can you sort it (in original or manipulated form) into point, linear, and areal categories? If not, perhaps a different sort of illustration would be more useful than a map.
- **Construction.** Don't assume that you will get it right the first time! Like writing, preparing illustrations takes time, reflection, and, usually, revision. Carefully select patterns, line widths, symbols, and type to ensure that the distinctions between different items are clear—neither too sharp nor too subtle. Ensure that all the information can be read with ease, and that type font sizes are appropriate both to the scale of the map and to the categories of information they identify. Include only the necessary level of detail, without distracting embellishments. Does the final product show the data to their best advantage? Are there any changes you could make to clarify an especially important point? Will the message in the illustration be self-evident and clear to every reader?
- **Checking.** Check your work carefully for completeness, accuracy, and readability. As with all of your work, be sure to proofread. Are locations, routes, boundaries, and other lines accurately placed? Are all the spellings correct? Are any numbers correct and complete? Are all the patterns correctly placed, and are they all readable? Is there a short, informative title? Is a written explanation necessary, in the text or under the illustration? Have you included scale and directional indicators? Have you credited your sources?
- **Final copy.** Once you are sure that your draft is correct, make a good copy for presentation with your written material. Remember to proofread the final copy before submitting it.
- **Identification.** Identify each map with a figure number (e.g., "Figure 2") and a cross-reference in the text.

Mapping with computers

Computers have added greatly to display and analytical capacities in geography and cartography, notably through *computer-assisted digital mapping* and *Geographic Information Systems* (GIS). Programs that incorporate increasingly sophisticated integrated analysis can be invaluable for exploring the relationships between different sets of data derived from the same area. Some remarkable interactive programs are available, and some statistical manipulation can now take place *inside* a computer mapping program. In many cases you no longer need to do the kind of pre-analysis data manipulation that students faced even in the recent past. Some software packages include ready-made maps of American states and Canadian provinces, counties, and census tracts, into which you can insert your data. Most basic packages are "user-friendly," but you need to be aware of their limitations, especially with respect to design possibilities and display effectiveness.

You also should be aware that automated cartography can be time-consuming. If you have just one map to produce and the assignment is relatively minor it may make sense to use traditional techniques. Still, as time goes by there is increasing pressure on higher-level students to prepare all their maps and graphs using a computer.

Computer-generated digital mapping and "vector" system GIS programs use "overlays" containing digitized data on different subjects to create maps showing the relationships among them: for example, field patterns, ground water availability, well sites, drainage systems, crops grown, seasonal livestock densities, and land values; or elevation, streams and rivers, soils, population distribution, political boundaries, roads, telephone and electricity lines, water and sewage pipes, and present and planned buildings. Individual overlays can be "peeled off" as necessary to reveal different types of interrelationships. Numerous practical applications of this approach are possible in the work world; hence "vector" system GIS is now an important part of the training of geographers and other environmental scientists. Some instructors have their students create personal websites on which they can display their GIS course assignments.

Another kind of GIS, the "raster" system, is an invaluable tool for analyzing geographic phenomena, but the mapped results may look coarse because of the grid structure into which data are placed.[3] This advanced system takes a long time to learn and most students will not use it.

Like any map, GIS shows points, lines, and areas—though in GIS jargon "areas" are generally referred to as "polygons." Two kinds of data are gathered: *attribute* data, which pertain to specific areas such as census tracts (and which often come packaged with map data), and *image* data, which can be scanned and digitized from satellite images, aerial photographs, or various other printed sources. If you are creating maps using a GIS program, be careful not to combine too much overlapping information. Clarity should be your goal.

Here are some basic questions to ask if you are considering using a computer-generated digital mapping program:

- Is your software package capable of doing what you wish?
- Are you familiar with the digital techniques? Do you have enough experience with them to complete the project within a reasonable time?
- Is your data set in a form that the software can read?
- Is the essential locational information (map boundaries, rivers, contours, etc.) available in a digital form that the software can read? Your library or department may have a collection of standard base maps and thematic maps in digital form. If not, is a suitable up-to-date base map available for digitizing?
- Have you been careful to differentiate between point, line, and polygon (area) data?
- Have the data in your database been given the coordinates required for use in an automated program? If not, is conversion software available to transform the existing digital data into the required format?
- If you are digitizing or extracting information from a map with a scale or projection different from your base, the fit will not be precise enough for accurate comparisons. However, there are ways of coping with this problem, so don't hesitate to ask your instructor for help.
- Whatever program you use, be careful to select patterns and densities that fully reveal what *you* want to display and reveal in the illustration. Don't use pre-set displays, just because they are there, if they are not right for your illustration.
- Check to see that all of your data (points, lines, patterns) are where they should be.
- Check the spelling on the map, as you would the text of a paper.
- Check to be sure that you have provided a key so the reader will know how to read your map.
- Make sure that the hard-copy will be of suitable quality for your intended purpose.

- If you are able to use colour, be aware that your mapped results will direct the viewer to think one way or another. Assume that you are to map the majority of percentage "yes" and "no" responses to a questionnaire as they pertain to a question posed for the people of region "x" and that you are able to plot the data by census tracts. If you chose a dark colour (e.g., red) for "yes" and a light colour (e.g., yellow) for "no," your viewer will be attracted to the "yes" pattern, whether or not it is in the majority of tracts. In other words, think carefully about what and how you want to convey the information, using colours. The same holds true for black and white patterns because more densely packed patterns may attract the eye more quickly than lightly packed patterns.

- Be careful how you mix colours on the page. For example, if you use blue and purple, or purple and red, you might mask the order of your data, and your viewer may find it difficult to identify quickly what is important. Use of pure colours might be better, such as blue and red. Be aware that some colours connote different things in different cultures. For instance, pink is feminine for most people in Canada, the U.S., and Britain, but not in Japan; green is a sacred colour for Muslims, and purple means prostitution in some Southwest Asian countries.[4] In short, know your intended audience because some colours may be inappropriate in certain circumstances.

Although a working acquaintance with computers is now an essential part of anyone's education, it is possible to become too caught up in the technology. You don't want to get so bogged down in technical problems that you lose sight of your primary goal: to portray spatial data.

GIS-based projects published on the web

You may be asked to develop GIS-based projects and reports for publication on your department's website. Your writing skills will apply for such a task, just as for other assignments. Don't use jargon; use straightforward English. Typically, such a report will be about 3,000 words, not including tables and figures. The following briefly outlines some guidelines:

- **Title page.** Start with a concise and informative title for the work. It should mention the study area and the fact that you are using GIS. Then identify yourself (and partner(s), if applicable). Finally, list the major sections of your work (using the headings below) and provide "links" to each of those sections.

- **Abstract.** (See Chapter 4.)
- **Introduction.** Start with a clear statement of purpose before discussing the project in greater length, placing it within the context of pertinent literature. Briefly state, at minimum, the following four main objectives:
  - Objective 1: To *identify* the factors pertinent to "whatever."
  - Objective 2: To *develop* a GIS-based model for identifying (or evaluating or determining) . . . .
  - Objective 3: To *apply* the GIS-based model to . . . .
  - Objective 4: To *evaluate* the strengths and limitations of the model.
- **Study Area.** Identify the study area, with a "link" to a map showing the location.
- **Research Approach.** Discuss the research procedures, how you developed or amended the GIS program to your specific needs. Identify the data used, its quality, and conversion techniques.
- **Research Findings.** Write your conclusions and draw the reader's attention to the various tables and figures (maps and graphs) that you will have developed, linking them so the reader can go back and forth with ease.
- **Conclusions.** Review what you have done and identify the conclusions you have reached. If appropriate, include recommendations for action, or further study. Identify the strengths and limitations of the study.
- **Literature Cited.** Fully identify all books, refereed journal articles, and reports you have cited within the body of your work. Do not include material that you have not cited.
- **Acknowledgements.** You also may wish to include an acknowledgements page to thank those who provided data or otherwise assisted you.

Each of the above items will have its own section on the website. Be sure to include "link" buttons for "sending" your reader to the figures, forwards and back, and include a "home" link or button to return the reader to the title page. Finally, as with all of your writing, check the spelling and grammar in the text and in the tables and figures, and double check that what you have written makes sense and is complete.

TABLES

Tables present data in a systematic manner, using rows and columns under appropriate headings. How do you decide whether a table is the best way to present your data? Ask yourself these questions:

- What is the purpose of the table? Could the data simply be listed in the text? Is time a variable? If so, perhaps a graph is in order, possibly in addition to a table. Is area or location significant in the distribution of the data? If so, perhaps a map would be more useful than a table.
- Do the data in the table complement—not duplicate—what is in the text? (Try to avoid repetition as much as possible.)
- What categories should the table show? Are there different ways you might group the data in the table? Which ways will most effectively convey the information and make the point you intend?

The following are essential for any table:

- **Table number and title.** The assigned number should appear at the top of the table and in the text at an appropriate place. The title should be brief and informative. Avoid abbreviations in titles; also remember that the noun is "percentage," not "per cent" (see Table 8.1 below).
- **Headings:** The headings for each row (reading across) and column or group of columns (reading down) must be clear; make sure that any abbreviations are explained either in the text or in a footnote. In Table 8.1, years, totals, and percentages appear along the top; employment categories at the left identify the data in each row.
- **Footnotes:** If you need to define or otherwise explain something, use a footnote, as in Table 8.1.
- **Source(s):** List the sources of your data.
- **Cross-reference:** Double-check to make sure you have referred to the table in the text (e.g., "See Table 4").

Table 8.1 presents fictional data for which percentages have been determined. Data from two time periods, and their proportional importance, are shown in an orderly manner. The table thus shows both absolute and relative data.

Table 8.1 Non-governmental labour force in Zania by industry, 1981 and 1991

Activity	1981		1991	
	Total	Percentage	Total	Percentage
Agriculture	101,649	16.5	104,343	15.7
Forestry and Fishing	20,000[a]	3.2	27,000[a]	4.1
Mining	16,940	2.8	19,648	2.9
Manufacturing	156,353	25.4	162,773	24.4
Construction	98,282	16.0	99,345	14.9
Commercial Services	196,643	32.0	226,783	34.0
Professional Services	24,980	4.1	26,862	4.0
TOTAL	615,147	100.0	666,754	100.0

[a] Estimated.

Source: Government of Zania, *Annual Labour Force Statistics* (Zunila: Bureau for Statistics, 1992), p. 10.

EQUATIONS

Place short and simple equations (such as "E = mc²") in the line of text, and long or complicated equations apart from the text, as in the two examples below. If you are presenting two or more equations, you can identify them with numbers (in parentheses) in the right margin, as in these examples. Unless an equation is standard—and thus is generally well known—you should explain any abbreviations in one of three places: immediately below the equation (example 1), in the equation itself (example 2), or in the text (example 2, for reference to "RF").

Example 1

The equation for deriving standard deviation is:

$$(\sigma) = \frac{\Sigma (x_i - \bar{x})^2}{n} \tag{1}$$

where σ (the Greek letter "sigma") = standard deviation;

Σ = sum of;

x_i = score of measurement, with i the value of observation;

$\bar{x}$ = mean of a group of scores or measurements; and

n = number.

Example 2

The equation for determining the vertical exaggeration on a topographic map is:

$$\text{Vertical exaggeration (VE)} = \frac{\text{Vertical scale (VS)}}{\text{Horizontal scale (HS)}} \tag{2}$$

Once this is noted, you can use abbreviations. The vertical and horizontal scales are expressed as representational fractions (RF).

Suppose that the HS is 1:50,000 and the VS is 1 inch to 400 feet. The VS expressed as an RF is 1:4,800.

$$\text{Therefore: VE} = \frac{1}{4,800} \div \frac{1}{50,000}$$

DIAGRAMS

Diagrams can help readers to visualize relationships and processes. Cross-referencing with the text is essential, using the appropriate figure numbers, to ensure that the relationship between any diagram and the ongoing text is clear. Any abbreviations should be explained either in the text or in a legend. Three examples of diagrams follow. Figure 8.6 shows, in outline form, a student's strategy for undertaking research on the environmental world view of Buryats in Siberia. This imaginative, visually appealing diagram makes it easy for readers to grasp the step-by-step process.

Figure 8.7 shows how the same student captured diagrammatically the relationships between constructs and themes explored in her work—relationships that would be difficult to explain in words alone. Note the use of different type sizes, symbols, and lines.

Figure 8.8 reveals linkages between various present-day factors that act in and on farms, and relates individual farms to others within a regional system. The end result is visually pleasing and instructive: it communicates essential information about various "forces" that impinge on an individual farm, identifies key factors within the farm, and indicates that effects and responses occur at various levels within the system. The use of different boxes, type,

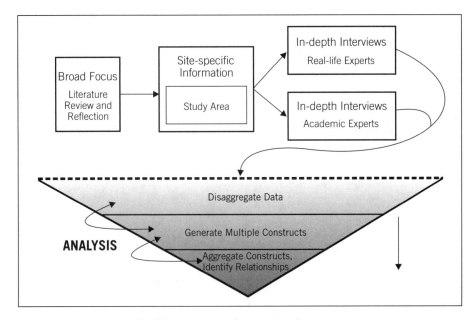

Figure 8.6 A graphic illustration of a research strategy

SOURCE: Kamilla Bahbahani, by permission.

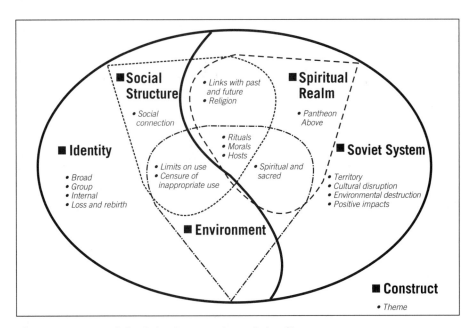

Figure 8.7 Model of the integration of the five constructs

SOURCE: Kamilla Bahbahani, by permission.

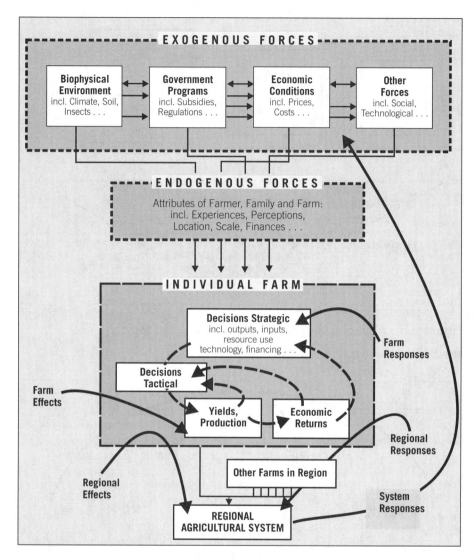

Figure 8.8　A conceptual model of agricultural adaptation to climate variation

SOURCE: Barry Smit, reprinted by permission of the author.

shading, and arrows creates an informative diagram that says, in effect, "read me." The descriptive title places the information in a broad context (agricultural adaptation to climate variation) and indicates that the diagram represents a conceptual model.

GRAPHS

A graph is useful for providing a visual image of the numerical relationships among several observations. Take your time when compiling a graph; think about the style as well as the content. Figure 8.9 shows two ways of presenting the same data: one "bad" and one "good." On the left, the "bad" example uses the kind of 3-D effect that is common in advertising and is often available in elementary computer graphing programs. Such illustrations may be appealing, but they are not suitable for academic work. The direct, face-on style and clear patterns of the "good" example make it easy to read and facilitate comparison of data.

There are several kinds of graphs:

- **Line graphs** reveal continuous change by using an actual or implied linking of data between observations (e.g., yearly measures of farmland devoted to grapes). As an example, Figure 8.10 shows the numbers of students employed within a jurisdiction (on the vertical or y-axis [ordinate]) over a period of 24 years (on the horizontal or x-axis [abscissa]). A variant of a line graph is a logarithmic graph, used when data values are too high to show on an arithmetic axis.
- **Bar graphs** consist of either vertical or horizontal columns; the length of each column is proportionally equal to the value being displayed. Two vertical types are used in Figure 8.9, each with 1981 and 1991 data, to demonstrate their usefulness for comparative purposes. Figure 8.9 uses separate columns for each category; the height of each column indicates the proportional value of the category it represents. Figure 8.11, in contrast, uses one column, the total of which is 100 per cent, so the relative importance of each category is shown within the column. A **histogram** is a type of bar graph often used to show frequency of observations over time. Figure 8.12 is an example of a **climagraph**, which uses a histogram in the lower half to show rainfall totals for each month in "Mothvale," and a line graph to show average monthly temperatures.
- **Pie graphs** or **proportional circles** are divided into wedges indicating proportions (Figure 8.13). They are used to emphasize the proportions of individual items making up a whole (e.g., amounts of land by usage, the ethnic components of a population, or the various fauna within a particular region). These graphs show at a glance the relative importance of each category.

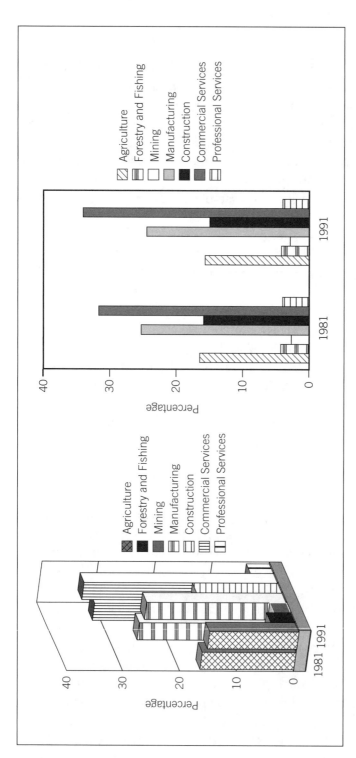

Figure 8.9 Two examples of bar graphs (a "bad" style on the left and a "good" one on the right), showing the non-government labour force in Zania by industry

SOURCE: Table 8.1 on p. 135.

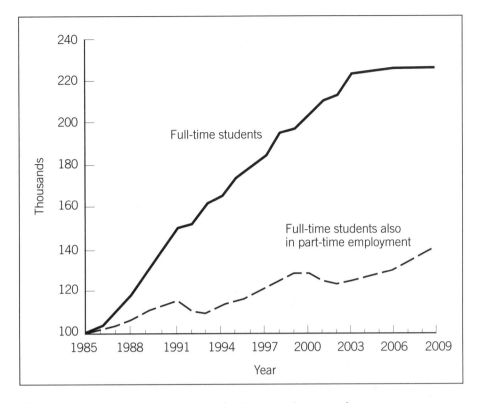

Figure 8.10   Full-time students also in part-time employment

SOURCE: Fictitious.

- **Scattergrams** show graphically the amount of correlation between two sets of statistical data. The vertical or y-axis generally depicts the dependent variable while the horizontal or x-axis depicts the independent variable. Figure 8.14 includes three scattergrams, each of which shows a different set of data, plotted according to the x- and y-axes. The graphs show that differences between the sets of data exist in terms of the strength and direction of relationships. The latter are shown by a line that identifies the goodness of fit to the regression equation, with r obtained by measuring deviations from the line computed by least squares (about which you will learn in methods courses). Graph 1 shows a strong positive relationship; graph 2 shows there is no relationship; and graph 3 shows there is a weak negative relationship.

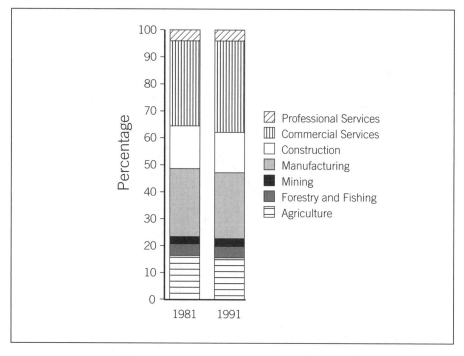

Figure 8.11 A subdivided 100 per cent bar graph, showing the non-government labour force in Zania by industry

SOURCE: Table 8.1 on p. 135.

- **Population pyramids** show population distribution by age categories—say, in five- or ten-year steps—with males on the left half and females on the right. Additional information can be added, as in Figure 8.15.

A basic consideration when you are deciding what kind of graph to use is the nature of the data to be shown. *Discrete data* (collected at a single time) normally are presented on bar or pie graphs, whereas *continuous data* (showing a sequence through time) generally are displayed as line graphs or histograms. Many computer programs now permit you to develop all sorts of graphs; often it is possible to change the form of your graph simply by changing a setting. Be attentive if you work with a coloured screen but will print your work in black and white; the finished product may look quite different once printed on paper. Don't be lured into using some gimmicky elements of a computer program. They may look good on the screen but they may not convey

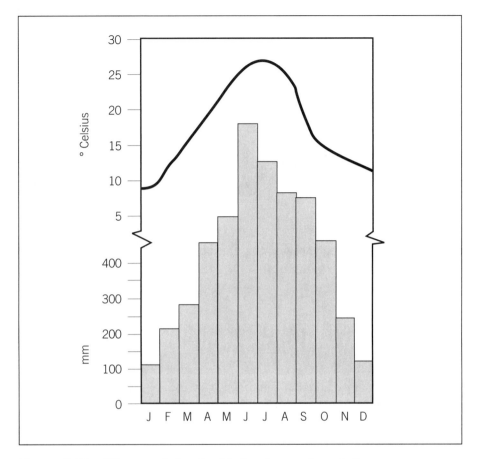

Figure 8.12 **Climagraph for the Mothvale weather station**

SOURCE: Fictitious.

the data clearly or accurately. Indeed, the program may even *distort* your data on a graph—just as can happen on a map. The result will be that your illustration would be a lie.

If you don't have access to a computer, you still can make acceptable graphs if you follow these guidelines:

- Use a ruler and black ink. Draw a rough version on graph paper and then trace it onto plain paper.
- Place your dependent variable (the one you measure) on the vertical or y-axis (ordinate) and your independent variable (the one you have manipulated) on the horizontal or x-axis (abscissa). The or-

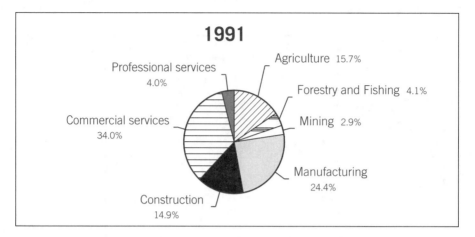

Figure 8.13 **Pie graph showing the non-government labour force in Zania by industry, in 1991**

SOURCE: Table 8.1 on p. 135.

dinate is normally about three-quarters the length of the abscissa, although there are exceptions to this rule.

- Label the axes clearly.
- Use large and distinctive symbols, with different line thicknesses and styles.
- Include only essential details (a cluttered graph is difficult to read).
- Include a legend to show what various symbols represent.
- Identify the source of the data.

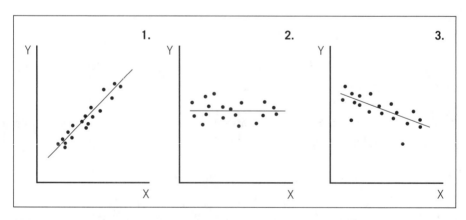

Figure 8.14 **Three scattergrams**

SOURCE: Fictitious.

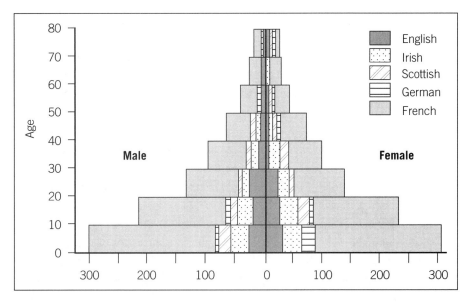

Figure 8.15 **Population characteristics of L'Orignal, Ontario, by sex, age, and ethnic origins, 1871**

SOURCE: Compiled by David B. Knight from data obtained from the manuscript copy of the 1871 Census of Canada.

- Give the graph a title and number, and refer to it by that number in your text (e.g., "See Figure 7").
- Ensure that your work is neat, clean, and clear, and that all your spelling is correct.

Printers

If you have an old printer, you may find that it can't produce graphic material to the standard required for your submissions (but some older printers have better gray-scale capability than newer ones, so check with your departmental technician). Laser and good inkjet printers can produce excellent quality coloured and monochrome illustrations. Remember, if you use colours, chose them with care; pretty colours do not necessarily result in good illustrations.

NOTES

1   See, for instance, A.H. Robinson, et al., *Elements of Cartography*, 6th ed. (London: Wiley, 1995); T.A. Slocum, et al., *Thematic Cartography and Geographic Visualization*, 2nd ed. (Englewood Cliff, NJ: Prentice Hall, 2005); B.D. Dent,

Cartography: Thematic Map Design, 5th ed. (New York: McGraw-Hill, 1999); M. Monmonier, *Mapping It Out: Expository Cartography for the Humanities and Social Sciences* (Chicago: University of Chicago Press, 1993); J.S. Keats, *Understanding Maps*, 2nd ed. (London: Longman, 1996); E. Tufte, *Visual Displays of Quantitative Information*, 2nd ed. (Cheshire, Conn.: Graphics Press, 2001); M.-J. Kraak and F. Ormeling, *Cartography: Visualization of Spatial Data*, 2nd ed. (Harlow, England: Longman, 2003); and M.W. Pearce, *Exploring Human Geography with Maps* (New York: Freeman, 2003). Some older books remain useful, including E. Raisz, *Principles of Cartography* (New York: McGraw-Hill, 1962); D. Greenhood, *Down to Earth: Mapping for Everybody*, rev. ed. (Chicago: University of Chicago Press, 1964); and G.C. Dickinson, *Statistical Mapping and the Presentation of Statistics* (London: Arnold, 1973).

2   M. Monmonier, *How to Lie with Maps* (Chicago: University of Chicago Press, 1996). See also his *Bushmanders & Bullwinkles: How Politicians Manipulate Electronic Maps and Census Data to Win Elections* (Chicago: University of Chicago Press, 2001), and J. Black, *Maps and Politics* (London: Reaktion Books, 2003). For the impact of new technologies and their consequences for society, see M. Monmonier, *Spying with Maps: Surveillance Technologies and the Future of Privacy* (Chicago: University of Chicago Press, 2002).

3   See, for example, P.A. Longley, et al., eds., *Geographic Information Systems and Science,* 2nd ed. (Chichester: Wiley, 2005); C.P. Lo and A.K.W. Yeung, *Concepts and Techniques of Geographic Information Systems*, 2nd ed. (Toronto: Pearson, 2007); C.A. Brewer, *Designing Better Maps: A Guide for GIS Users* (Redlands, CA: ESRI Press, 2005); and K-t. Chang, *Introduction to Geographic Information Systems,* 5th ed. (New York: McGraw-Hill, 2009).

4   J. Krygier and D. Wood, *Making Maps: A Visual Guide to Map Design for GIS* (New York: Guilford, 2005), p. 267.

CHAPTER 9

Doing Field Work and Writing about It

In all environmental disciplines, the environment itself is a source of information. Doing field work is fundamentally important. Your focus will depend on your discipline and the nature of the course you are taking. For example, attention may be directed to the physical or biological properties and processes in a particular place or region; to the spatial patterns of human-made features or processes; to the interplay between human and natural processes; or to human understanding of a particular place and its people and the way they interact. These broad statements don't encompass all that you will do as you seek to make "sense of that vast panorama of fact and fiction, pattern and event, on the surface of the earth," but they are a general guide.[1] The topics for investigation are diverse, and huge in number, but the information obtained by field work will provide you with primary data. It takes time to learn how to "read" and "understand" what you "see," thus the need for field courses in which you will learn a variety of techniques.

Two types of writing (with illustrations) are associated with field work: field notes (which may or may not be evaluated by the instructor) and assignments (which are submitted for evaluation). Following some pointers on preparing for field work, this chapter focuses on keeping a field notebook; making sketch maps, sketches, and photographs; keeping a field diary; constructing a questionnaire; and writing a formal field report based on analysis of your data.

PREPARATION

Statement of purpose

You will want to know what you are going to do, hence—likely with guidance from the instructor—you will identify the purpose of the field trip and what is to happen. Reading and some initial map work may be required, and perhaps also some initial data collection and analysis, before you can develop a clear statement of purpose (i.e., unless one is provided for you) and what exactly you plan to do. If you are to do soils work, for example, you will want to identify the parameters of your research design and the methodology to be used. Depending on what you learn as you prepare, you will know what you will have to take along on the trip, including extra clothes, food, and equipment.

Equipment

The basic items needed for most types of field work are a bound notebook (with both lined and graph paper), a clipboard (with a plastic cover in case of rain), several pens and pencils, paper clips, and a sturdy small shoulder bag (the kind you can get in outdoor or army surplus stores). Other items will depend on the weather, the season, and the location where the field work is to be carried out: a whistle, boots or sturdy waterproof shoes, sunglasses (with UV protection), a first-aid kit, insect repellent, a hat, a raincoat or winter jacket, and so on.

A hand-held (or palm) computer notebook can be useful for recording data in the field, and a laptop computer is invaluable for recording and analyzing data immediately, either in the field or at home in the evening. You also may need maps (e.g., topographical, soils, land-use, and geological), air photos if available, a multi-metre tape measure (usually one per group is enough), a calculator, and, if you don't have a computer with you, a ruler and a compass for constructing graphs. A directional compass is essential in certain types of terrain and for certain types of field assignments. Depending on the instructor's directions, other special instruments (such as soil augers, wind gauges, thermometers, tree corers, etc.) may be needed as well.

If you are to use a Geographical Positioning System (GPS)—a powerful geo-referencing tool—you will need an accurate sketch map to verify any digital analysis of your data that you may do at a later time. To use a GPS effectively when recording field data, you will have to know the location of set benchmarks, their coordinates and elevation, and, in some instances, the satellite forecast before you set out.

Safety

When doing any kind of field work it's wise to establish a "buddy" system; each set of "buddies" should have at least one whistle, in case of emergencies. If you are to be away from a settlement for an extended period, you should give someone in authority a plan (similar to a flight plan or itinerary) of your anticipated route and check-in time. Additional measures are essential if you are going to conduct research in a remote or isolated location. If you are undertaking upper-level field studies—independently or in a group—your department may require that you first complete a risk assessment.

Permissions

To do certain types of field work it is necessary to obtain permission. If you are to apply a questionnaire or otherwise engage in direct contact with people, it is increasingly common, even for undergraduate students, that the questionnaire and research methods (e.g., the sampling method) must be approved by an ethics committee. Check with your professor to be sure. If you will be doing work that involves other people, or have to walk through someone's property, it may be necessary to get prior permission, perhaps by a phone call but preferably by letter. In any event, be sure you have a letter of introduction from your professor to show if the occasion arises.

Finally, remember that while you are in the field you are a representative of your institution: always be polite to people and respectful of property. It would be a pity if some smart comment or questionable behaviour on your part turned anyone against later generations of students. In a rural area it's wise to let the local police know what you are doing. Your instructor may do this, and may also arrange to have an article appear in the local newspaper to inform local residents that they can expect to see strangers in their midst.

FIELD NOTES

Good field notes are not for your eyes only: they must be legible and understandable to others—including your instructor. This is critical if your field research is part of a larger project concerned with change over time at a particular location, since future researchers may require your data. It is essential to record your observations in an orderly fashion, whatever the length of the field exercise—two hours, a day, or several weeks. *Specific* "systematic" observations, recorded on the spot, make up the first part of field notes; the second part consists of later reflections on those observations ("summary statements"). Systematic observations can be recorded in point form, whereas summary statements usually are more formal and written in complete sentences. If you are to make observations at several sites, they need to be standardized, so it's important to prepare your format in advance. That way you can see easily whether you have obtained all the required information before you move on. Similarly, if you are using classifications, you should work them out (as far as is practical) in advance; if using abbreviations or symbols, place summary lists or tables in your field notebook for ready reference.

The data you gather in the field may later be mapped using a computer-assisted cartography program, perhaps as part of a Geographic Information System (GIS) study, or used in some other fashion, but first they must be recorded in the systematic observations section of your field notebook (or in a hand-held computer notebook).

Systematic observations

FIRST THINGS FIRST

Your instructor may give you a handout outlining the purpose of the field assignment; if so, keep it with your field notes (perhaps taped or glued into your notebook). At a minimum, record at the top of the first page the basic premise of the study, a statement of purpose, any basic questions or hypotheses, and an outline of the methods to be used. These set the stage for everything that follows. What should you do if your instructor doesn't provide a question, but just tells you to "investigate" a particular problem? First, always seek to explain an observed pattern; don't simply describe it. Never be satisfied with merely writing down whatever data you gather, unless the instructor has explicitly stated that nothing more is expected. Second, to help you do this, take the problem statement provided by the instructor and either reformulate it into a central question or advance a hypothesis; this will serve as a guide to the types of information you must gather. Third, with those requirements in mind, consider the types of research methods you will need to use. Together, your basic question or hypothesis and research methods will help you determine both what to include and what to exclude from your field observations and data collection.

FORMAT, TIME, LOCATIONS, AND HEADINGS

At the start of each day, write down the date and time at the top of a new page. Record the name of anyone who assists you in gathering data. You may find that each instructor has a particular format for recording field notes, but the following schema is typical. Divide the paper into two sections: a narrow column on the left—perhaps 6 to 8 centimetres (2.5 to 3 inches) wide—for locational and time references and headings for different types of observations (e.g., "soils," "slope," "land use," and—if these might influence the data and therefore be needed later to explain any possible discrepancies—"weather conditions" such as precipitation, wind direction, and cloud cover); and a wider column, taking up the rest of the page, for recording your observations.

Whenever you note an observation, also record the time and location. The latter will be particularly important later, when you are analyzing the day's findings far away from the site, since data may have been gathered from numerous locations. Map referencing to the coordinates of the topographical sheet is best.

References to "near the maple tree" (how near?) or "20 metres west of the farm house" are of no use, unless the locations of the maple tree and the farm house are recorded, too. Using a GPS will make it easy to be precise.

Headings, as noted above, can be useful for keeping your observations discretely organized on the page, especially when you are recording several types of data at each site. At first it may be advisable to take more summary, point-form notes than you think are really necessary. In time, as you gain experience in determining what is important and what is not, your field notes will become increasingly useful.

SELECTIVITY

You need to be selective in deciding what observations to record in your notebook. Either the instructions you receive from the course instructor or the purpose that you and your colleagues have established will help you determine what is important. For example, if you are to read the landscape for evidence of particular forms of architecture, you may not need to record observations about soil types—unless, for instance, (a) you have hypothesized that there is a relationship between soil type and architectural form, or (b) you know that a record of such data could be useful in the future, for another purpose. If, on the other hand, you are to observe existing recreational land uses in order to determine the potential for further recreational development, you will need to seek and record a wide variety of data, including soils; slope; existing land uses; population density, distribution, and pattern; recreational activities; settlement types and spread, and economic base; governmental institutions; and so on. In short, the nature of the problem will help you define what you need to observe, measure, and record. Before beginning to collect your data, you also should have an idea of the types of analysis and presentation that will be required at a later time. These requirements may dictate your sampling methodology and will help you identify any additional information that you may need to record, such as the distance from a benchmark.

CONSISTENT AND ACCURATE RECORDING

It is important that you record *all* your raw field observations in a consistent manner first, before grouping them into categories, adding interpretations, or drawing conclusions. If you record the data one way at one site and another way at the next, they may be useless. At the outset, then, consider the exact nature of the data to be collected, and then record them consistently, accurately, clearly, and completely. If you use equipment to record data, make sure the settings and units on the instrument(s) are appropriate to the task, and that the settings are taken from the pertinent instructional manuals.

ANALYSIS, TABLES, AND GRAPHS

In some cases it may be appropriate to present your findings in tables for each site for which you have data; in others it may not be possible to put the information into tabular form until all the field work has been done. The data tables you develop then will be available for further analysis and display using various methods. For instance, if you are gathering data on truck travel flows at a particular intersection, you might summarize the findings by time periods on tables before graphing and mapping them. All three methods can be useful, and much of the work can be done in your field notebook, though entering your data into a computer spreadsheet will save a lot of time and effort.

Not all the information you record will be numerical; subjective data can be essential for certain types of field work (especially in some cultural and social geography courses). Such data can be more challenging to work with than simple, objective numbers, but they can be rewarding. What you want to do is find patterns or linkages among observations that at first glance may seem completely unrelated. To do that, you need to experiment. Sort the data into different categories—and don't get stuck on the first ones that come to mind. You may need to "manipulate" the data in some way before you can draw any inferences from them. Thus the data carefully recorded in your field notebook from each observation, at each site, and (depending on the nature of the study) perhaps between sites, may become "useful" to you only later, once you have analyzed the total data set using any of a wide range of statistical and quantitative techniques. If such analysis is necessary, be sure to include the results in your field notebook.

SKETCH MAPS

The old saying that "a picture can speak a thousand words" applies to field work. Two types of "pictures" are important for field notes: sketch maps and drawings of the scene under consideration. If the scale is not critical—if all you need is the approximate pattern of distribution—then a sketch in map form may be adequate. Such sketch maps are easy to do, since they are only approximate recordings of what you see. If more accurate information is needed, you will want to use graph paper and a tape measure. In cases where you don't have time to take accurate measurements, there is an alternative: simply count the paces it takes you to walk the distance you want to measure. For practice, lay out an exact course of 15 metres (approximately 49 feet) or more, pace it forwards and backwards (to make sure you've counted correctly), and divide the length by the number of paces. Walk in a relaxed but steady manner; don't take unnaturally long strides. This will give your average pace (from instep to instep or, if you prefer, from heel mark to heel mark). Thereafter, all you need to do is pace out the line to be measured (e.g., the perimeter of a property, or a building's dimensions) and multiply the number of paces with the known length of your pace. On any sketch map you make, remember to include a scale, an informational key, and a directional indicator (north arrow).

SKETCHES

Even if you think you can't draw, with practice your sketches will become quite adequate to record some essential information that relates directly to the field project.

In fact, sketches generally are more useful than photographs, since you can omit some of the things you see that are not essential to the problem at hand. The scale of the observations will help you determine how much detail to include. Imagine, for instance, that you are to sketch a soil profile of the area 1 metre by 4 metres (approximately 3 feet by 13 feet): a proportional scale (say, 10 centimetres to 1 metre, or perhaps 3 inches to 1 yard) will have to be determined so that your sketch will reflect accurately what you see and measure. The components of the profile should be labelled. Or, perhaps you have to draw a particular form of joint in a log building; here again, exact measurement may be important. In a sketch of a large area, by contrast, it would be impossible to record everything you see. An artist sketching the scene would need to be selective; so do you. How?

Ask yourself why a drawing would be useful. The answer will depend on the purpose of the field work. Are you studying particular rural or urban land uses, drainage, valley slope, vegetation patterns, architectural forms, river bank erosion, upland ecosystems, plant colonization and succession, glacial land forms, garbage dumps, or quaternary riverine patterns? What particular things are you examining? Would a drawing show its outline, distribution, or pattern to advantage? What should you include? What should you emphasize? What can you omit? When sketching a valley from a nearby hilltop, for example, you may want to record the vegetation (such as wind breaks) or field patterns in some detail, to the exclusion or downplaying of buildings. If you are to draw a river meander, the location of above-water gravel in the river will be more important than trees that are well away from the bank. Just as you think about what to record in words, think about what might be useful to record in a drawing or a sketch map.

Whereas some of your sketches may suggest merely approximate scale and measurements, others will need to be accurate. For instance, a sketch or block-diagram (essentially a relief model) of a woodlot edged by an old rock wall can be approximate (within reason), but a cross-section derived from topographical maps must accurately reflect the scale and contour intervals shown in those printed sources. Sketches of equipment you use in the field must also be accurate in detail, especially if proportional relationships are important, or if a dial or other means of measurement is recorded. Finally, accuracy obviously is important if you are asked to produce a detailed sketch of, for instance, the precise placement of equipment in a field study site.

PHOTOGRAPHS

Photographs can be invaluable for certain types of field work. You will need to ensure that you are close enough to isolate the subject—unless you intend to show it in relation to other features in the landscape—and that the details you want are clearly visible. If the scale is not evident, place an item (such as a lens cover or a marked measuring stick) in the scene, or beside the subject of the photo. The Polaroid camera is a boon to many researchers, since it makes a colour image available almost immediately for insertion in your field notebook, with space beside it for comments on the contents. A digital camera permits you to edit the image at a later time and transfer it to a program such as PowerPoint, or, if you have the proper equipment and supplies, to download and print photographs on-site. Some cameras permit you to record the date and time on the photograph. Remember to include in your field notebook a reference to each photograph taken. Never digitally manipulate a photograph with the intent to deceive; to do so would be unethical. As David Griffen notes, "To change a photo—beyond minor tweaks to correct color balance or remove red-eye, for example—is to undermine the truth."[2]

Summary statements

Periodically you will want to summarize your observations. Summarizing sessions during the course of the day may be possible, but at a minimum a late-afternoon or evening session is needed. Here are three suggestions:

- After gathering data at one site and recording them in your notebook, you may be able to take a few minutes to reflect on what you have been doing. If you have questions that can't be answered readily, jot them down for later reference. Also, take the time to note any overarching observations you may have made about what you have been doing. These insights may be useful later.
- A lunch break can be a good time for interim summarizing, collating data, or, more importantly, reflecting on what you have written

in your notebook during the morning. Are there questions that come to mind once you step back from the actual data-gathering?

- The most important time for summarizing is late afternoon or evening; don't put this task off until the next day. Reflect on what you have done during the day and jot down your major observations while they're still fresh in your mind. You will be pleasantly surprised later, when these observations help you write your field report. Your summary observations may include anything from specific insights to general feelings about what you have seen and recorded. The evening is a good time to compile summary tables of the data gathered; then you will have a summary of the day's data ready either for immediate analysis or for later compilation and analysis with data from other days.

General comments

It's important to keep your field notes tidy, since they may be examined and graded by your instructor. But field notes can serve other purposes too. Clearly written systematic observations, with the pertinent data, sketch maps and drawings, and sound summarizing statements, will be invaluable, whether for class assignments—such as lab work using your data, or research papers based on your field work—or in studying for an examination. Without care, accuracy, and consistency, however, your field notes will be of questionable value. Be sure to think in advance about what you plan to record, why, and how you will do it; check all your field notes the same day you take them, to make sure no important data are missing; and, finally, make sure that everything you write is legible.

Field diary

In addition to field notes, some researchers find it useful to keep a separate *field diary* or *journal* for general thoughts and reflections. In contrast to a field notebook, with its step-by-step, detailed record of observations, a diary is more reflective, recording your musings, ideas, insights, questions, tentative conclusions, and possible hypotheses that may be worth further consideration. One way to start is with a succinct summary—no more than a sentence or two—of your activities and general observations from that day's work. Then stand back from those observations and allow yourself to think more broadly and abstractly than you can when you're preoccupied with the nitty-gritty of field work. For some types of field work in human geography, recording your emotional responses to the day's activities can also be useful. Leave space at the end of each day's entry so that you can return to it and add further reflections, perhaps days or even weeks later.

PREPARING A QUESTIONNAIRE

Writing—i.e., designing—a questionnaire is not something that can be hurried. Knowing your purpose for using a questionnaire, and the population to be sampled, will help you design a document that is unambiguous and, thus, easily understood. A well-designed questionnaire should provide you with useful primary data. There are many types of questionnaires, as you will learn in a methods course. Here are several fundamental points to help you as you proceed (assuming you have received any necessary approval from your departmental or university ethics committee to administer your questionnaire, and have fulfilled their expectations of your department or university):[3]

- You may be required to obtain a signed, "informed consent" letter from the people you intend to speak with, and to leave each of your respondents with a copy of it, prior to proceeding with your questionnaire.
- At the top of your questionnaire, list the project's title, your name, and the names of your department and institution. You might also include a contact phone number or email address, especially if the questionnaire is to be left for respondents to complete on their own.
- By way of introduction, as briefly and clearly as possible, identify the project's purpose and state precisely what you expect of your respondents.
- Keep the questionnaire focused. Don't include superfluous material or unnecessary questions that will distract the respondents.
- Avoid ambiguity. Be precise and clear when you pose a question.
- Avoid bias. Select your words (and sentence structure) with care to avoid influencing the respondents' reactions to what you pose.
- Get the attention of your respondents by putting factual questions first. More complex material can be placed later.
- Think about the order of your questions. A logical flow should be evident, which will help your respondents. You may use headings or brief explanatory comments to indicate any change in the line of inquiry.
- Be tidy. If your questionnaire is messy (e.g., with spelling errors, mixed fonts, inadequate spacing) or is poorly organized, you may lose the attention of your respondents.
- Conclude with a "thank you."

REPORTING FIELD ASSIGNMENTS

When field notes are to form the basis of a more formal report, your instructor may give you the headings required to organize your paper. Note that requirements regarding format, quality of sketches and mapping, terminology, and so on in geography differ from those in biology; physical geography generally calls for a different style than human geography. Whatever the style, organize your thoughts according to the instructions given in class; ignoring them may seriously affect your grade. If, on the other hand, your instructor simply tells you to "write a report on the field assignment," with no elaboration, you shouldn't take this as permission to write a rambling, unstructured report. Always present your observations and findings in a clear, succinct, systematic, logical manner. What should go into your report? Consider the following suggestions.

1. **Title.** A clear, descriptive title will leave no question about what you are reporting. For instance, "Contemporary Commercial Land Uses in Windsor, Louisiana" is preferable to "A Pretty Neat Place for Today's Shopping: Variations in the Commercial Patterns of a Small Southern Village"; a simple "Leda Clays in the Ottawa Valley" is better than "Clays that Subside and Flow when They Get Wet in the Ottawa Valley Region." If a scientific name is applicable—say, for a plant study—use it so that your reader knows at once what the report is about. Also, be sure to identify yourself, your course, and your instructor on the title page.

2. **Introduction.** Your introduction should identify, in general terms, the purpose of the field study, its basic premise, and the essential question you sought to answer or the hypothesis you sought to test (all these will have been determined *before* the field work was done). The introduction is also the place to situate your study within a broader context. The latter may be given to you by your instructor, but if you have to establish the context yourself, you may need to refer to the course readings, or perhaps a lecture. As has been observed elsewhere in this book, all research has a context, so you must indicate what it is. In addition, you should indicate where and when the field research was carried out; it may also be important to state why that particular place was chosen for study.

3. **Research methods.** Identify what research methods you used and why. If, in the course of the research, you encountered problems and had to make modifications, note these too.

4. **Data gathered.** Present your findings in a straightforward manner, without interpretation. It may be appropriate to present some data in tables, graphs, or maps as well as in your text.

5. **Data analysis.** Identify the methods you used to analyze the data, whether qualitative, quantitative, general cartographic, or GIS. State what statistical tests you applied and note the results. Again, data may be presented in tables, graphs, and maps. You need not include all the data or workings of the analysis, although an appendix of that material may be attached. In any event, be sure to keep your detailed analysis of the data—if only so that, later on, you can see where you might have gone wrong.

6. **Discussion.** Identify the observations you have drawn from the data you have gathered and analyzed. Are there any patterns and processes that became obvious only after you worked with the raw data? What are they? How do they relate to the basic purpose of your study? Do they provide a reasonable answer to your essential question? Do they support your hypothesis? If not, explain why. What unanswered questions remain? Were there any problems with the data, or perhaps with your analysis? Spell these out in your report.

7. **Conclusion.** State your conclusion. Succinctly declare whether or not your original purpose has been fulfilled. Finally, *don't* include any new material.

NOTES

1   Anne Buttimer, *The Practise of Geography* (London: Longman, 1983), p. 12.

2   David Griffen, "A Question of Trust," in Chris Johns, ed., *National Geographic Guide to Digital Photography* (Washington, D.C.: National Geographic Society, 2006), p. 16.

3   For additional suggestions, see G. Bridge, "Questionnaire Surveys," in *The Student's Companion to Geography*, 2nd ed., eds. A. Rogers and H.A. Viles, 230–41 (Oxford: Blackwell, 2003).

CHAPTER 10

Writing a Proposal,
Research Paper, and Thesis

All writing requires planning, but when you are to tackle a major work, such as an honours research paper (sometimes called an undergraduate thesis) or a graduate thesis, the planning stage is formalized. You may have to prepare a proposal to be approved by one or more faculty members before you will be permitted to do the necessary research. In addition, some senior undergraduate courses require the preparation of a research proposal as an exercise. And, of course, a report of some kind is required once any research has been done. The structure of research papers and theses generally conforms to standards accepted across all disciplines. This chapter highlights the essential elements of proposals, research papers, and theses.

Before proceeding, it is important to remember that your work may well involve **ethical issues** (see Chapter 2). Though not repeated in this chapter, they must be be taken into account when you prepare a proposal, or undertake research that will result in a research paper or thesis.

Two general comments: first, always double space your work, unless otherwise instructed, and, second, never put anything on the back side of any page in any work you prepare.

PREPARING A PROPOSAL

Sometimes referred to as a research design, plan, or prospectus, a research proposal can be thought of as a strategic plan for a research project. It should logically and systematically identify *what* you propose to do and *how* you propose to do it. Since the development of a research proposal is a cyclical process, you may revise it as often as necessary until it defines the research project that you wish to do and believe you can complete. Figure 10.1 illustrates the typical, major steps in developing a research proposal; the double-headed arrows indicate the non-linearity of the process. Often, particularly when you are revising your proposal, it is helpful to work backwards, in other words, begin with the data collection tools, and then move to the objectives and to the statement of the problem. Doing so enables you to check for consistency throughout your proposal.

Questions you need to ask	Steps you need to take	Selected elements of each step
• What is the problem or question that intrigues me (or that was assigned to me)? • Why should it be studied?	Select and analyze the research problem; formally state the research question or problem	• Identify the nature of the problem and its context (background information). • Define the study area. • Analyze the problem, the factors that affect it, and the relationships among the elements of the problem. • Decide on the focus and scope of your research. • Justify the research (provide a rationale).
• What information is available? • How does my topic relate to others' work? • What is excluded from my research? • What contributions might my research make?	Prepare the literature review	• What published literature is available? (What is already known about the problem?) • What other sources of information need to be considered? Identify individuals, groups and organizations, and unpublished information sources (e.g., reports, records and computer databases).
• Why do I want to conduct this research? (What is the purpose of this research?) • What do I hope to achieve with the study results?	Formulate the research objectives	• What are my general and specific objectives? • What assumptions am I making? • What questions will be asked? (Or, what hypotheses might be developed?)

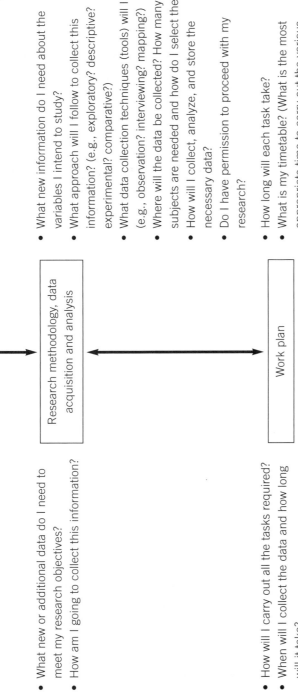

- What new or additional data do I need to meet my research objectives?
- How am I going to collect this information?

Research methodology, data acquisition and analysis

- What new information do I need about the variables I intend to study?
- What approach will I follow to collect this information? (e.g., exploratory? descriptive? experimental? comparative?)
- What data collection techniques (tools) will I use? (e.g., observation? interviewing? mapping?)
- Where will the data be collected? How many subjects are needed and how do I select them?
- How will I collect, analyze, and store the necessary data?
- Do I have permission to proceed with my research?

Work plan

- How will I carry out all the tasks required?
- When will I collect the data and how long will it take?
- What resources (e.g., equipment, transportation) do I need, and what resources do I have, to carry out the project?

- How long will each task take?
- What is my timetable? (What is the most appropriate time to carry out the various research tasks?)
- What is my tentative chapter outline?
- Is my bibliography complete and accurate?

Figure 10.1 Typical steps in developing a research process

Although your preparation will become increasingly comprehensive and focused, preparing to write a proposal is similar to preparing to write an essay. First you will have to identify the topic, limit it to something workable, determine how to relate it to others' work, know what data you will need, decide what methods of data collection and analysis will be appropriate, and suggest what the outcome of the study might be. You need to lead your reader to the inevitable conclusion that your topic has never been tackled before, or at least not from the perspective you propose.

There are several parts to any proposal:

- **Statement of research problem.** A formal statement of your research problem includes a description of the nature of the problem (e.g., its size, distribution, and severity) and a rationale for conducting the research. The formal statement also needs to **define your study area**. Even if your instructor has pre-selected the study area, you still need to define it and explain why it is an appropriate choice for study of the problem at hand. Similarly, you need to indicate your **approach or viewpoint** in your statement. There are two parts to this section. First, you need to suggest the broader context (disciplinary or interdisciplinary) into which your specific study fits, and the significance of its contribution. For instance, you might indicate that the data from your planned research will help to fill a gap in available information about the topic. Second, you need to indicate the philosophical stance that you intend to take (e.g., will you use a postmodernist approach? will you take a feminist perspective?).

- **Limitations**. Either within the statement of research problem or as a separate entry, identify the limitations of your study. Know in advance what you will *not* do! In addition to specifying areas or issues that you will not research, you need to identify such practical limitations as constraints on time or funds available for the research, field access difficulties, and other challenges that may affect your ability to conduct your proposed research.

- **Literature review.** The relationship between your proposed work and the existing work in the field must be established. You need to convince your faculty advisers that you are familiar with at least the general research background within which your study fits. For undergraduate students, the literature review need not be lengthy, but it should establish that you are aware of the significant current thinking in the subject area and that you have thought about the contributions your research might make to the topic.

- **Research objectives**. The objectives of a research project summarize what you hope to achieve by your study and how the results could be used. Both general and specific objectives need to be specified; general objectives indicate what you expect to achieve in general terms, while specific objectives identify the various aspects of your problem and specify what you will do in your research, where, and for what purpose. In terms of **anticipated results**, depending on the nature of your research, it may be appropriate to comment on the areas of application to which you will direct your possible conclusion (e.g., "Findings will be interpreted, in part, in relation to their significance to planning in the rural region.").
- **Research methodology and data acquisition.** This section should indicate succinctly what information/data you will gather, where and when you will do so, the methods by which you will collect your data, and your reasons for choosing those methods (interviews, questionnaires, experiments, mapping, remote sensing, archival material, instrumentation, etc.). Any sampling procedure you propose should be justified in terms of both the material or population to be sampled and the sampling method to be used.
- **Data analysis.** This section should explain how the information you have acquired is to be evaluated, organized, and analyzed or synthesized. The amount of data and the basic question and/or the nature of the hypothesis will determine how the data ought to be handled. If you will be dealing with large quantities of data that will require computer processing and the use of statistical techniques, these should be identified—and you should be able to defend your choice of techniques. Make sure, as part of your methodology, that you have secured any necessary permissions to proceed with your research; for example, ensure you have permission from the owners to access your selected site, or have obtained written approval of the ethics board at your institution.
- **Work plan.** A work plan or schedule summarizes the different components of your research project and identifies how and when you will perform all the tasks associated with your project. Schedule the dates on which each task should be started and completed. Indicate how long it will take to do the various sections of the assignment. Be realistic—you're likely to need more time than you think. You can develop a **tentative chapter outline**, and identify the anticipated logical order of the sections for the proposed written presentation.

- **Bibliography.** List the most important works you consulted in developing your proposal. Include only the works you cite (and perhaps discuss) in the proposal. Don't "pad" by providing a long list of books, chapters, and journal articles that may relate to the topic, but that you have not yet read.

Identifying Sections

Check with your faculty adviser about departmental style preferences. Many departments want sections to be identified with subheadings and/or letters or numbers (see pp. 77, 165, 166).

RESEARCH PAPERS AND THESES

A research paper prepared during the final year of undergraduate study (or a thesis completed as part of graduate studies requirements) is a long, formally structured essay based on the analysis of some chosen set of data or set of propositions. Findings may be based on primary research, but secondary sources may also be important.

In general, a research paper focuses sharply on a particular issue, whether a body of literature or some specific field or lab research. A thesis is expected to show a reasonable degree of originality in approach, data or principle, and technique or method; the originality required in a thesis is greater than in a research paper but somewhat less than in a doctoral dissertation. Whether your analysis is qualitative or quantitative in nature, the quality of your text always is important.

Your department may have guidelines for the outline and content of research papers or theses. The guidelines for research papers may be similar to those for essays (see Chapter 5), but, as with theses, the structure generally does not allow for much flexibility (see below). Typically, a draft copy of a research paper or thesis is, given to your supervisor and your "committee" (composed of other faculty members) for their evaluative comments, after which you will have the opportunity to make changes—from minor corrections in grammar and spelling to major revisions that may require rewriting some chapters or even (dread the thought) the whole work. The final paper then is submitted for examination. The grade is usually based on evaluations of the written work together with an oral defence.

Plan how you will spend your time: doing background reading; formulating your proposal; reading additional material; gathering data; analyzing data; planning the written report; writing the first draft; revising and redrafting;

preparing and checking illustrations, submitting the work to your instructor or supervisor (and committee) for evaluation; and (if such is permitted or required) revising as necessary before submitting the final product. To do all this within a period of perhaps only two or three terms (or semesters), you will have to work out a rough schedule in advance. Remember, too, that you should build in some time for yourself—and be prepared to take short breaks when you need them.

Your paper must be structured in a logical manner. Since it will have numerous paragraphs, headings of some sort are not only helpful but required. There are two systems for headings: either written headings/subheadings (A and B in Table 10.1) or some form of numerical differentiation, whether Roman numerals, alphanumerics, or decimals (C, D, and E in Table 10.1). Appropriate headings and subheadings are used with each of the latter three methods (none are shown in the table).

Headings

Headings normally have a ranking, from most to least important, as in the examples in Table 10.1. Whichever method you choose (or are told to use), be sure to employ equivalent symbols for sections of equivalent importance. And, once you have settled on a style, follow it consistently.

The structure of your work follows a progression, from the title of the project at the start to the references or bibliography at the end. In between is the substance of your work and its supporting apparatus. The outline of your work will have five or sometimes six components, organized as follows:

PRELIMINARY MATERIAL

Numbering the preliminary pages: The preliminary pages (which include the title page, abstract, acknowledgements, list of tables, and list of figures) are numbered with small Roman numerals (e.g., ii, iii, iv) located at the bottom centre of each; however, the number for the title page, which is "i," should not appear. The appropriate Roman numeral for each item in the preliminary pages appears in the table of contents. The number of the first page for each chapter also appears there.

Title page: The title page identifies the title of your work; your name, department or program, and institution; a statement (centred, near the bottom of the page) such as "In partial requirement for the B.Sc. [or whatever] degree in the Department of *Y* at the University of *Z*"; and the year.

Table 10.1 Systems for headings

Written Headings and Subheadings

A.

TITLE OF PROJECT [Centred]

SECTION HEADING [lst order; flush left, all upper case]

Subheading [2nd order; flush left, bold]

<u>Subheading</u> [3rd order; flush left, underlined]

Subheading [4th order; flush left, not underlined or bold]

B.

TITLE OF PROJECT [Centred]

<u>Section heading</u> [lst order; centred, underlined]

Subheading [2nd order; centred, not underlined or bold]

<u>Subheading</u> [3rd order; flush left, underlined or bold]

Subheading [4th order; flush left, not underlined or bold]

<u>Subheading</u> [5th order; indented start to a paragraph, underlined]

Numerical Headings and Subheadings

C. *Roman Numeral*	D. *Alphanumeric*	E. *Decimal**
I.	A.	1.0
A.	1.	1.1
1.	a.	1.11
2.	b.	1.12
B.	2.	1.2
II.	B.	2.0

* An alternate decimal system places all the numbers at the left margin.

Abstract: A short summary of your thesis, the abstract is placed on a separate page, following the title page. Mention the basic purpose, methods, and major findings. (For suggestions on how to write the abstract, see Chapter 4).

Table of Contents: On another separate page, list the following items, with the spacing as noted, and with appropriate page numbers located to the right of each heading:

Table of Contents

Abstract
Acknowledgements
List of Tables
List of Figures

CHAPTER
1.   Title
2.   Title
etc.

Appendices
1.   Title
etc.
Bibliography/References

Acknowledgements: Acknowledge any agency—governmental or not—and your university or department, that financially or otherwise supported your research. Unless they have requested otherwise, remember to name and thank everyone who assisted you, including your supervisor. If some people did not want to be identified by name, you could still thank them in a generic way, by saying something like " . . . and those who chose to remain anonymous."

Lists of Tables and of Figures: Include the number of each item, its title, exactly as it appears within the text, and its page number. These lists should appear on one page if each list is short; otherwise each list appears on a separate page, to accommodate the length of the lists. See below under *Illustrations* for the several types of material that are grouped under the term "figure."

Chapter titles: The list of chapter titles should *not* include subtitles if you are writing a research paper. However, a fully detailed table of contents—including all subtitles—can be useful, and may even be required, in a graduate thesis. If you are unsure of the level of detail to include, it's a good plan to ask your supervisor.

INTRODUCTION
Your introductory chapter should outline the following, usually with separate headings:

- precise statement of purpose
- definition of terms
- hypothesis/hypotheses, objectives, basic question
- study area; i.e., the location (in a lab or a field site) where you did the work
- scope
- methodology
- limitations (what the work will *not* do)
- organization of the work (what each chapter contains, to guide the reader)

SUBSTANTIVE REPORT (PRESENTED IN SEVERAL CHAPTERS)

- extended review of literature
- data: sources and reliability; method(s) of collection and analysis
- presentation of data (which may include illustrations—see below)
- analysis of data (ditto)
- summary of findings

CONCLUSION

- conclusions and, when appropriate, recommendations.

Do not introduce new material in your conclusion.

APPENDICES
Some studies include appendices. Material that supports your discussion in a particular section of the text but is somewhat tangential to its main thrust may be better placed in an appendix. Data too extensive to be included in tables within the main text also may be listed in an appendix. Each appendix is num-

bered, given a title, and begins on a new page. Be sure to refer to any appendices in the body of your text—e.g., "(see Appendix 3)"—so the reader knows to look for it at the end of the work. Appendices always appear before the bibliography.

BIBLIOGRAPHY OR LIST OF REFERENCES

As in all major works, the final item you have to attend to is the list of books, book chapters, journals, and other items that you have *cited* in your study. Works you may have read in the course of your research but have chosen not to cite are not included. Assembling and checking the nitty-gritty details in your bibliographic list will be a frustrating and time-consuming chore if you didn't record full publication information when you first read the material (see Chapter 5).

ILLUSTRATIONS

Most research papers and theses are illustrated. An illustration, whether graph, table, map, or photograph, generally appears on the page immediately following the one in which it is identified in the text. It is not acceptable to use the back of the previous page to print your illustrations. If an illustration takes up more than half a page, no text—other than the illustration's title, any very brief comments made about the illustration, and, most important, source information—should appear on the page with it. *Figures* include maps, graphs, and photographs. Each is numbered, either from 1 to *n*—i.e., from the beginning to the end of the main text (any figures in appendices are numbered separately, within each appendix)—or, using the decimal system, from beginning to end of each chapter (in which case Fig. 2.1 would refer to Figure 1 in Chapter 2). *Tables* are numbered in a similar fashion. Always refer to your figures and tables in the text, using their identifiers (Figure 5.1; Table 4.3) so the reader knows when to turn to them. And remember to discuss each illustration in your text; otherwise the reader will wonder why it is there. Finally, remember to assign page numbers to all illustrations, in sequence with your text page numbers.

STANDARDS

Just as you are required to meet the highest standards in the conduct of your research, so, too, your writing standards must be high. Aim for a clear and precise style, with proper attention to the fundamentals of grammar, spelling, punctuation, and logical structure.

CHAPTER 11

Quotations and Documentation

QUOTATIONS

Quotations can add authority to your writing as well as help you avoid charges of plagiarism. But you should use them with care: never quote a passage just because it sounds impressive. Be sure that it really adds to your discussion, either by expressing an idea with special force or cogency, or by giving substance to a debatable point.

Guidelines for incorporating quotations

1. Integrate the quotation so that it makes sense in the context of your discussion and fits grammatically into a sentence:

   ✗ Whether Bill Gates is a visionary is debatable. "640K ought to be enough for anybody" is now highly ironic.

   ✓ Whether Bill Gates is a visionary is debatable. His 1981 prediction that "640K ought to be enough for anybody" is now considered laughably ironic.

2. Be accurate. Reproduce the exact wording, punctuation, and spelling of the original, including any errors. You can acknowledge a mistake by inserting the Latin word *sic* in brackets (see p. 244) after it. If you want to italicize or underline part of the quotation for emphasis, add either *my emphasis* or *emphasis added* in brackets at the end. Note the placement of the period, inside the quotation marks but outside the brackets:

   "The group's findings were *not* confirmed [my emphasis]."

3. Include as part of your text, enclosed in quotation marks,

   • no more than three or four lines of prose:

     As Brian Graham has observed, ". . . identity is defined by a multiplicity of often conflicting and variable criteria."[1]

- no more than four lines of verse. Use a slash (/) to indicate the end of a line:

  Patrick Duffy, discussing Ireland's "north," quoted from a poem by John Hewitt: "The hospitable Irish / come out to see who passes / bid you sit by the fire / till it is time for mass."[2]

4. A long quotation from a text—normally four or more lines—is usually single-spaced and introduced by a colon. If the first line of your block quotation is the first sentence of a new paragraph, indent the first line; otherwise, start flush with the left margin:

   The environmental determinist Ellen Semple (1863–1932), using a rather florid style, wrote that nature moulded humans:

   > Man is a product of the earth's surface. This means not merely that he is a child of the earth, dust of her dust; but that the earth has mothered him, fed him, set him tasks, directed his thoughts, confronted him with difficulties that have strengthened his body and sharpened his wits, given him problems of navigation or irrigation, and at the same time whispered hints for their solution.[3]

   For long quotations from poems—four or more lines—you should write the words line for line as originally written.

5. For a quotation within a quotation, use single quotation marks:

   Our reporter got this comment from the forester: "I shouted 'timber' as I tried to get out of the way of the falling tree but I stumbled. The tree fell on me, and I screamed 'help' to my mates."

6. If you want to omit something from the original, use an ellipsis (see p. 250). To omit a full line of a poem, use a full line of single-spaced periods:

   Cedar and jagged fir

   . . . .

   Against the gray

   and cloud-piled sky.

7. If you want to insert an explanatory comment of your own into a quotation, enclose it in brackets:

> "The presidents of the three universities [Minnetonka, St. Paul, and Mankato] strenuously objected to the government's proposed funding scheme for universities."

DOCUMENTATION

In any work you do, it is essential that you fully and accurately record the source of all material—ideas, quotations, diagrams, graphs, maps, etc.—that is not your own. This documentation will give support and substance to your writing and will permit your readers to locate original source material.

There is no one "right" method of documentation: not only do different disciplines follow different conventions, but individual instructors have different preferences and expectations. One may tell you to follow the style of a particular journal; another may direct you to a specific style guide. If not, choose one of the systems outlined below. Whatever you do, remember to follow the one "golden rule" for documentation: *always be consistent.* No matter what system you choose, use it consistently throughout any piece of work: *never* change systems in midstream.

Basic characteristics

Among the essential elements that are normally included in all systems of documentation are the following (their order may vary slightly):

- Surname of the author(s).
- First name and initials of the author(s); the scientific method uses only initials. Do *not* include titles (such as Colonel, Sergeant, Sister, or Rev.), affiliations, or degrees that precede or follow names. *Do* include any prefix (as in "*von* Schmidt" or "*van der* Lankfeld") that is an essential part of a name, or any final element appearing at the end of a name, preceded by a comma (as in "Blair Mannerly, *Jr.*" or "Angus George, *II*").
- For a book: the work's full title, including any subtitle, and if applicable the name of the editor(s) or translator(s), preceded or followed by "ed." or "trans."
- If applicable, the number of the volume or, if you have used a complete work in several volumes, the total number of volumes.

- For a book: the place of publication, the name of the publisher, and the year of publication. Note that the names of some publishers can be condensed (e.g., "Basil Blackwell" to "Blackwell," or "Jamieson and Sons, Ltd.," to Jamieson), whereas the names of associations and university presses are normally given in full (but see p. 179).
- For a journal: the title of the article and the name of the journal. Do *not* include the name of the company that published the journal (e.g., Sage).
- For a journal: volume number; issue number, (in some cases only if issues are paginated separately); year; and the pages where the article appears.
- For online journal, magazine, and newspaper articles: the same information as for their print counterparts and, in some cases, the complete URL (or DOI) and date of retrieval.
- For original content from online sources: the name of the site and/or the organization that maintains the site; the title of the document (if one is present), and in some cases the complete URL (or DOI) and date of retrieval.

Styles

While most disciplines in the physical, biological, and social sciences now use (A) the *parenthetical-reference* style, some use (B) the *notes-bibliography* style or (C) the *consecutive listing* method. Again, details vary between and even within disciplines.

Because documentation styles vary so widely, this section provides guidelines, not firm rules: check with your instructor or department to make sure you are following the preferred practice. The basic characteristics of the three styles (A, B, and C) are outlined below, with examples. The essential elements required for documenting cartographic, archival, and architectural records are outlined at the end of section C.

A. PARENTHETICAL REFERENCES

This method is followed, with variations, by the Council of Science Editors (CSE), the American Psychological Association (APA), and the Modern Language Association of America (MLA).[4] In this "bare-bones" approach to referencing, citations are given to the authorities/sources used to support/test the development of the case being argued, without any tangential discussions or references. This style has two main features:

- Brief references (termed *citations*) are located at appropriate points in the text.
- At the end of a work, a section entitled *Reference List, References,* or *Works Cited* includes the full publication information for those works *directly referred* to in the text; other works that you may have consulted but have not referred to directly in the text are not listed or are placed in a separate list called Works Consulted or Bibliography.

In 2006 the CSE published the seventh edition of its manual, *Scientific Style and Format: The CSE Manual for Authors, Editors, and Publishers.* This revised edition details three variations of the CSE style: the name-year system, the citation-sequence system, and the citation-name system. The examples that follow illustrate the name-year citation system. Should you require additional information about the citation-sequence or citation-name systems, your school library will have a copy of the *Scientific Style and Format,* 7th edition.

In the text

1. The citation usually is placed at the end of the sentence in which you refer to the work, with the author's surname and, in most cases, the year of publication in parentheses (note the presence or absence of a comma in the citation); the MLA style does not include the year:

   CSE   . . . tourism in India's national parks (Hannam 2005).

   APA   . . . tourism in India's national parks (Hannam, 2005).

   MLA   . . . tourism in India's national parks (Hannam).

2. Sometimes the citation is provided within a sentence:

   CSE, APA   Metaj (2007) reviewed the protected area system.

   (or)   In 2007, Metaj reviewed the protected area system.

   MLA   Metaj reviewed the protected area system.

If you are citing a work that has no declared author, you should give the title or a shortened version of it of the work. (e.g., *Cognitive Mapping and What*

Children See may be shortened to *Cognitive Mapping*). For a well-known literary work that you are citing repeatedly, you may want to abbreviate using only the initials (e.g., in a study of the geographical dimensions of Coleridge's "The Rime of the Ancient Mariner" you might use *TRAM* or *RAM* as your abbreviation).

3. If a citation is to an entire work, and the page reference is not important, all you need to include is the year. For MLA no citation is required.

> Thaler and Plowright (2006) support this hypothesis.

If, however, the citation is to something specific within the source and not just to the work in general, then you must include the specific reference. When referring to a specific page or table, include the number (note that in the CSE style there is no comma between the author's name and the year, and no period after *p*). Note that the APA style uses an ampersand (*&*) within the citation (but not in the text), whereas the CSE and MLA styles use *and*:

CSE   (Smyth 2006, p 121–123)
      (Jones 2008, Table 1)
      (Boertmann 2008, fig 1)
      (Gray and Wyly 2007, chap 17)

APA   (Smyth, 2006, pp. 121–123)
      (Jones, 2008, Table 1)
      (Boertmann, 2008, fig. 1)
      (Gray & Wyly, 2007, Chapter 17)

MLA   (Smyth 121)
      (Jones, Table 3)
      (Boertmann, fig. 7)
      (Gray and Wyly, ch. 17)

4. Citing a work with multiple authors can be tricky.[5] When shortening a list of authors, use "et al.":

CSE   *Two authors:*

> Schlinkerheim and Smyth (2005, 2006) are investigating the areal spread of coyotes into eastern North America.

Three or more authors:

Abbott et al. (2007, p 298) are correct.

APA   *Two authors:*

Schlinkerheim and Smyth (2005, 2006) are investigating the areal spread of coyotes into eastern North America.

When there are three, four, or five authors, the APA cites all authors the first time a work is identified (example *a*), and in subsequent references only the first name and "et al." are used (example *b*) as well as the year if it is the first citation of a reference within a paragraph (example *c*):

(a)   Rafferty, Simon, Singh, and Ngamakwa (2005) are correct.

(b)   Rafferty et al. are correct.

(c)   Rafferty et al. (2005) are correct.

When there are six or more authors, cite only the surname of the first author followed by "et al.":

Abbott et al. (2007) focus on community fisheries management in the Zambezi River floodplains.

MLA   When there are more than three authors, list all names or give the first author's name followed by "et al.":

Stewart, Howell, Draper, Yackel and Tivy considered the implications of Arctic sea ice for cruise tourism.

Stewart et al. are correct.

5. When more than one work is cited within a single set of parentheses, the items are separated by semicolons. In the CSE style they are listed in chronological order (example *a*), but in the APA style they are listed in alphabetical order (example *b*):

(a)   (Kollman and Prakesh 2001; Wilkie et al. 2006; Bailey 2007)

(b)   (Bailey 2007; Kollman and Prakesh 2001; Wilkie et al. 2006)

6. If there are two authors with the same last name, include their initials. The CSE style does not use periods or spaces between two or more initials:

CSE   (Smith DG 2007; Smith LC 2006)

APA   (Jonah, A., 2002; Jonah, E., 2004)
        (Smith, E., 2005; Smith, J. M. , 2007)

MLA   (A. Best; C. Best)
        (S. Bailey; T. Bailey)

7. When citing two or more works by the same author(s), list the surname(s) once, with the dates following. If there are multiple publications for any one year, use a letter marker after the year to distinguish them:

CSE   (Gold and Gold 2005, 2007).

      (Le Billon 2005a, 2005b, 2006, 2007, 2008).

8. When citing a work that you have not read directly but know through a quotation in someone else's work, you should cite the source of your information about the work; do not cite the original.

CSE   According to Cameron (1887, cited in Merante 2001, p 25), . . .

9. Lecture or seminar notes can be cited as "personal communications." Because your notes do not contain recoverable (verifiable) data, personal communications can be cited in your text only, and not in your reference list. You could say, for instance:

In his lecture Professor Draper outlined ... (D. Draper, GEOG 529 lecture, November 10, 2009).

In the reference section

Although specific details vary depending on the subject (see the sample entries below), all subjects in the alphabetical group share some features of reference style in common. Here are the general guidelines:

1. List your references alphabetically by author: surname first, followed by given names or initials; do not number them.
2. For a work with two or more authors, list all their names. In some styles only the first author's name is reversed (Zhang, C., X. Guo); in others all the names are reversed (Zhang, C., Guo, X.).
3. When citing more than one work by a particular author in the CSE and APA styles, list the entries in chronological order (the date should usually follow the author's name); for works published in the same year, add a let-

ter marker to each date (2008a, 2008b, etc.). In the MLA style, multiple works by the same author are listed alphabetically by title, and after the first entry the author's name is replaced by three hyphens followed by a period.

4. For references to journals, always include the volume number and inclusive page numbers.

Sample entries

Several variations exist. Note the differences in punctuation (such as the absence of periods after initials in names in the CSE style), parentheses, italicization, presence or absence of page references, and so on.

JOURNAL

The issue number within a volume is included only when the volume is not paginated continuously, except for MLA style, where the issue number is always given. The CSE style abbreviates the names of journals and does not underline or italicize them. Also note the placement of initials, the treatment of *and* (some styles spell it out and others use an ampersand, but the CSE omits it). In addition, there can be variations in punctuation (colon or comma), italicization, and spacing. Note too the location of the year of publication. The APA style italicizes, as a unit, the journal name and the volume number. In the MLA style authors' full names generally are cited, whereas in the other styles only initials are given.

> *CSE*   Hugenholtz CH, Wolfe SA. 2006. Morphodynamics and climate controls of two aeolian blowouts on the northern Great Plains, Canada. Earth Surf Process Landf 31:1540–47.

> *APA*   Hugenholtz, C. H. , & Wolfe, S.A. (2006). Morphodynamics and climate controls of two aeolian blowouts on the northern Great Plains, Canada. *Earth Surface Processes and Landforms, 31,* 1540–1547.

> *MLA*   Hugenholtz, Christopher H., and Stephen A. Wolfe. "Morphodynamics and climate controls of two aeolian blowouts on the northern Great Plains, Canada." *Earth Surface Processes and Landforms* 31 (2006): 1540–47. Print.

JOURNAL, MULTIPLE AUTHORS

CSE   Wheaton E, Kulshreshtha S, Wittrock V, Koshida, G. 2008. Dry times: Hard lessons from the Canadian drought of 2001 and 2002. The Canadian Geographer 52:241–62.

APA   Wheaton, E., Kulshreshtha, S., Wittrock, V., & Koshida, G. (2008). Dry Times: Hard lessons from the Canadian drought of 2001 and 2002. *The Canadian Geographer, 52,* 241–262.

MLA   Wheaton, Elaine, Suren Kulshrestha, Virginia Wittrock, and Grace Koshida. "Dry times: Hard Lessons from the Canadian Drought of 2001 and 2002." *Canadian Geographer,* 52 (2008): 241–62. Print.

BOOK

Note the placement of *p* or *pp*: *p 10–20* (CSE) and *pp. 10–20* (APA). For a whole chapter in a book, refer to the pertinent pages (e.g., "26–41," not "Chapter 2"). In the CSE style, *80 p* is the total length of a book; although this is optional it can provide useful information to the reader. The terms *Publisher (Publ.), Inc.,* or *Co.* and *Press* are generally omitted. The names of university presses are written out in the APA style but abbreviated in the CSE and MLA styles. Note that when there are more than three authors the MLA and CSE styles allow you to list the first author only, followed by "et al.":

CSE   de Blij HJ, Muller PO, Williams RS, et al. 2005. Physical geography: the global environment. Canadian edition. Don Mills: Oxford Univ Pr. 756 p.

APA   de Blij, H. J. , Muller, P.O., Williams, R. S. , Conrad, C. T. , & Long, P. (2005). *Physical geography: The global environment.* (Canadian ed.). Don Mills, Canada: Oxford University Press.

MLA   de Blij, Harm J., et al. *Physical Geography: The Global Environment.* Canadian ed. Don Mills: Oxford UP, 2005. Print.

In the CSE style, if the place of publication is not well known, include more information, such as the name of the state. Note that the APA style always includes the state or country. The MLA style does not require the state, province, or country.

CSE   Primack RB. 2006. Essentials of conservation biology. Sunderland, CT: Sinauer. 535 p.

APA   Primack, R. B. (2006). *Essentials of conservation biology.* Sunderland, CT: Sinauer.

MLA   Primack, R.B. *Essentials of Conservation Biology.* Sunderland: Sinauer, 2006. Print.

BOOK WITH EDITION NUMBER

CSE   Bone, RM. 2009. The Canadian north: issues and challenges. 3rd ed. Don Mills, ON: Oxford. 320 p.

APA   Bone, R. M. (2009). *The Canadian north: Issues and challenges* (3rd ed.). Don Mills, Canada: Oxford University Press.

MLA   Bone, Robert M. *The Canadian North: Issues and Challenges.* 3rd ed. Don Mills: Oxford, 2009. Print.

BOOK, TWO AUTHORS

CSE   MacEachern A, Turkel WJ. 2009. Method and meaning in Canadian environmental history. Scarborough, Canada: Nelson Education. 352 p.

APA   MacEachern, A., & Turkel, W. J. (2009). *Method and meaning in Canadian environmental history.* Scarborough, Canada: Nelson.

MLA   MacEachern Alan, and William J. Turkel. *Method and Meaning in Canadian Environmental History.* Scarborough: Nelson, 2009. Print.

BOOK, EDITED

CSE   Ali SH, editor. 2007. Peace parks: conservation and conflict resolution. Cambridge, MA: The MIT Pr. 406 p.

APA   Ali, S. H. (Ed.). (2007). *Peace parks: Conservation and conflict resolution.* Cambridge, MA: The Massachusetts Institute of Technology Press.

MLA   Ali, Saleem H., ed. *Peace Parks: Conservation and Conflict Resolution.* Cambridge MIT P, 2007. Print.

BOOK PUBLISHED IN TWO LOCATIONS BY TWO DIFFERENT PRESSES

CSE   Rum D, Barley JV, editors. 1998. The geography of drink across an international border. London: Mugs; New York: Saloon. 356 p.

MLA   Rum, Derek, and Julia V. Barley, eds. *The Geography of Drink Across an International Border.* London: Mugs; New York: Saloon, 1998. Print.

The APA style does not address co-publication.

CHAPTER IN AN EDITED BOOK

CSE   Sprague JB. 2007. Great wet north? Canada's myth of water abundance. In: Bakker K, editor. Eau Canada: the future of Canada's water. Vancouver: UBC Pr. pp. 23–35.

APA   Sprague, J. B. (2007). Great wet north? Canada's myth of water abundance. In K. Bakker (Ed.), *Eau Canada: The future of Canada's water* (pp. 23–35). Vancouver, Canada: University of British Columbia Press.

MLA   Sprague, John B. "Great Wet North? Canada's Myth of Water Abundance." *Eau Canada: The Future of Canada's Water.* Ed. Karen Bakker. Vancouver: UBC P, 2007. 23–35. Print.

GOVERNMENT REPORT

APA   Parks Canada and the Canadian Parks Council. (2008). *Principles and guidelines for ecological restoration in Canada's protected natural areas* (Catalogue No. R62-401/2008E). Ottawa, Canada: Queen's Printer.

MLA   Parks Canada and the Canadian Parks Council. *Principles and Guidelines for Ecological Restoration in Canada's Protected Natural Areas.* Ottawa: Queen's Printer, 2008. Print.

ARTICLE IN AN ENCYCLOPEDIA

CSE   Mower A. 2005. Cutting grass. In: The American encyclopedia of gardening. Volume 1. Des Moines, IA. Roots; pp. 307–8.

APA   Mower, A. (2005). Cutting grass. In *The American encyclopedia of gardening* (Vol. 1, pp. 307–308). Des Moines, IA. Roots Press.

MLA   Mower, Alex. "Cutting Grass." *The American Encyclopedia of Gardening*. Des Moines: Roots, 2005. Print.

Unpublished Dissertation

CSE   Galloway, JM. 2006. Postglacial climate and vegetation change in the Seymour-Belize Inlet complex, central coastal British Columbia, Canada [dissertation]. Ottawa (ON): Carleton University. 157 p.

APA   Galloway, J. M. (2006). *Postglacial climate and vegetation change in the Seymour-Belize Inlet complex, central coastal British Columbia, Canada.* (Unpublished doctoral dissertation). Carleton University, Ottawa, Canada.

MLA   Galloway, Jennifer M. "Postglacial climate and vegetation change in the Seymour-Belize Inlet complex, central coastal British Columbia, Canada." Diss, Carleton U, 2006. Print.

Presentation

CSE   Dawson, JD, Steward, EJ, Scott, D. 2007. Climate change vulnerability of the polar bear viewing industry in Churchill Manitoba. Working paper presented at the Tourism and Global Change in Polar Regions Conference; Oulu, Finland.

APA   Dawson, J. D., Steward, E. J., & Scott, D. (2007, November). *Climate change vulnerability of the polar bear viewing industry in Churchill Manitoba.* Working paper presented at the Tourism and Global Change in Polar Regions Conference, Oulu, Finland.

MLA   Dawson, Jackie, Emma Steward, and Daniel Scott. "Climate Change Vulnerability of the Polar Bear Viewing Industry in Churchill Manitoba." Tourism and Global Change in Polar Regions Conference. Oulu, Finland. 27 Nov. 2007. Presentation.

Article in an Online Journal

When referencing electronic journals follow the format of their print counterparts.

As the following examples illustrate, the MLA style includes the date of access but not the URL.

The APA style includes the DOI (digital object identifier) wherever possible. If none is available, give the URL. No retrieval date is required.:

CSE   Singh S. 2008. Contesting moralities: the politics of wildlife trade in Laos. JPE [Internet]. [cited 2008 Sept 5]; 15: 1–20. Available from: http://jpe.library.arizona.edu/Volume15/Volume_15.html.

APA   Singh, S. (2008). Contesting moralities: The politics of wildlife trade in Laos. *Journal of Political Ecology, 15*, 1–20. Retrieved from http://jpe.library.arizona.edu/Volume15/Volume_15.html.

APA   Perez, L., & Dragicevic, S. (2010). Modeling mountain pine beetle infestation with an agent-based approach at two spatial scales. *Environmental Modelling & Software, 25*(2), 223-236. doi:10.1016/j.envsoft.2009.08.04

MLA   Singh, Sarinda. "Contesting Moralities: The Politics of Wildlife Trade in Laos." *Journal of Political Ecology* 15 (2008): 1–20. Web. 6 Oct. 2009.

WEBSITE

CSE   National pollutant release inventory. [Internet]. 2009. Ottawa: Environment Canada; [cited 2009, Nov]. Available from: http://www.ec.gc.ca/inrp-npri/default.asp

APA   Environment Canada. (2009, July). *National Pollutant Release Inventory.* Retrieved from www.ec.gc.ca/inrp-npri/default.asp

MLA   "National Pollutant Release Inventory." *Environment Canada*, 29 July 2009. Web. 10 Nov. 2009.

Using Notes with Parenthetical References

Both the APA style and the MLA style permit the use of notes to supplement the information contained in parenthetical references.

In the APA style, notes are used for two purposes:

- Content footnotes amplify substantive information.

- Copyright permission footnotes acknowledge the source of copy-righted material that has been reproduced or adapted.

In the MLA style, notes are used for two purposes:

- Content notes provide additional information or comment.
- Bibliographic notes provide evaluative comments on sources or supply information on references containing numerous citations.

When using notes, place a superscript Arabic numeral at the appropriate place in the text and insert the note after a matching numeral at the bottom of the page (footnote) or at the end of the text (endnote).

Notes should be used sparingly, since they divert the reader's attention from the main text. If you find yourself creating long, complicated notes in large numbers, it is probably better to use the notes-and-bibliography method of documentation described in section B.

B. NOTES AND BIBLIOGRAPHY

This system uses notes (either *footnotes* or *endnotes*) and a *bibliography*. It permits greater freedom than the parenthetical-reference style does, since it lets you use *bibliographic* and *content* notes to expand upon the material presented in the main text. When using notes, you must also include a full *bibliography* (see pp. 189–91).

Notes

Notes allow the reader to check sources, verify information, and consider points that may not warrant inclusion in the main text. Never put essential information into a note; if it is important to your argument, you should include it within the main body of the paper. Notes are useful for developing an argument (especially when the word count is limited), presenting commentary on your sources, expanding on a point made in the text, identifying specific sources, and so on. But including too many of them can interrupt your reader's progress. Try to include in the text as much of the necessary information as you can to keep the note itself as short as possible. To avoid clutter, remember that you don't need references to common knowledge or undisputed facts. Just be sure to provide source information in a note if you quote or summarize other people's ideas; such acknowledgement will help you to avoid charges of plagiarism.

Use footnotes/endnotes:

- to identify the sources of particular words or ideas;
- to provide additional relevant information, subsidiary discussion, amplification, or qualification of points made in the main text;
- to point out cross-references within your own work; and
- to acknowledge assistance received from people or agencies who, for instance, provided information, or gave you permission to use their data or their lab facilities, or undertake interviews on their premises, using their staff or clients.

Both footnotes and endnotes are identified with superscript numbers (i.e., numbers slightly raised above the line of words) in the text. *Footnotes* appear at the bottom of each page. *Endnotes* appear at the end of the paper, usually under the heading *NOTES*. The format for footnotes and endnotes is identical; only their locations differ.

Format

1. For the reader's convenience, you may place your footnote at the bottom of the page on which the citation appears. Be sure to leave enough space for it. You should also leave a quadruple space between the text and the footnote, to make the division clear. A line (either about ten spaces in length, or right across the page) between the text and footnote will help to make the division clear. If you are using a computer, the 'Insert' menu will usually have an option that allows you to easily insert and format both footnotes and endnotes.
2. Footnotes should be indented the same number of spaces that you use for a new paragraph. In student writing they should be single-spaced, in contrast to the double-spaced text. Leave a double space between entries.
3. If you choose to place all your notes at the end of your paper they become *endnotes*. Using a separate page, underline or italicize your centred title— *Endnotes* or simply *Notes*—and list the notes by number. Leave a triple space between the title and the first entry, and a double space between entries.
4. Whichever format you choose, remember to number your notes consecutively, using arabic numerals, and to put the corresponding number at the appropriate place in the text (usually at the end of the sentence in which

you make each reference), using superscript. The numbers should follow all end punctuation.

5. Some instructors permit the use of smaller font sizes for notes (e.g., 10-point) than the standard text size (e.g., 12-point). Find out what your instructor prefers before you prepare and submit your paper.

The following examples of documentation are restricted to the most common kinds of notes and bibliographic references. They are based on *The Chicago Manual of Style*.

Notes for first references

ARTICLE IN A JOURNAL

The abbreviation *p*. (or *pp.*, for multiple pages) may be used to indicate page numbers, but increasingly it is omitted from references. Most page numbers may be contracted to avoid unnecessary repetition (for example, *33–7, 465–82, 1277–9*); however, numbers between *10* and *20* should always be written in full (*10–11, 12–15, 214–18, 1342–1480*). If the issue of the journal in which the article appears is one of several bound together to form a single volume with *continuous* page numbers, you may leave out the week or month or issue number of publication—see the two examples—and just give the volume number, year (in parentheses), and page numbers. Here are examples of two styles:

1. Mujib Ur-Rehman and Nick Chisholm, "Factors Affecting Households' Participation in the Natural Resource Management Activities in District Abbottabad, NWFP, Pakistan: A Multivariate Analysis," *Journal of Asian and African Studies* vol. 42, no. 6 (2007), pp. 495–516.

2. Sumanta Bagchi and C. Mishra, "Living with Large Carnivores: Predation on Livestock by the Snow Leopard (*Uncia uncia*)," *Journal of Zoology* 268 (2006), 217–24. doi:10.1111/j.1469-7998.2005.00030.x.

BOOK BY ONE AUTHOR

Capitalize the first letter in the title and subtitle, as well as the first letters of all words except for articles, prepositions, and conjunctions. The author's full name, not just initials, can be cited. Terms such as *editor, compiler*, and *translator* are abbreviated. When the publisher's name is long, you should use its familiar short form; thus *Macmillan of Canada Ltd.* becomes simply

Macmillan.

   3. Sharon Beder, *Environmental Principles and Policies: An Interdisciplinary Introduction* (Stirling: Earthscan, 2006).

BOOK BY TWO AUTHORS (WITH PAGE REFERENCES)

Although the title page may list more than one place of publication (say, Toronto and Buffalo), you need only name the first place in your reference:

   4. Moses Brackish and Marshy Puddle, *Crossing the Dead Sea* (Toronto: Shore, 2009), 65–7.

BOOK BY THREE OR MORE AUTHORS

In a note only, the name of the first author is included, followed by "et al." with no intervening comma; in the bibliography, the first ten authors are listed, followed by "et al."

   5. Adrian Slater et al., *Plant Biotechnology: The Genetic Manipulation of Plants*, 2nd ed. (Oxford, OUP, 2008).

BOOK WITH ONE EDITOR

   6. Nicole Aubertin, ed., *Wind Tunnels* (Hamilton: Blowhard, 2009).

BOOK WITH TWO EDITORS

   7. Rick Riewe and Jill Oakes, eds., *Climate Change: Linking Traditional and Scientific Knowledge* (Winnipeg: Aboriginal Issues Press, 2006).

BOOK BY ONE AUTHOR AND EDITED BY ANOTHER

   8. Mark C. Serreze and Roger G. Barry, *The Arctic Climate System*, ed. J.T. Houghton, M.J. Rycroft and A.J. Dressler (Cambridge: Oxford University Press, 2005).

BOOK BY ONE AUTHOR TRANSLATED BY ANOTHER

You need not list the author's name when it is part of the title:

   9. *Plato's Republic*, trans. Meron J. Mueller (Sydney: Harbour, 2001).

Multivolume Work

> 10. I.M. Profound, *Philosophizing Geography*, 2 vols (Chicago: Hillock, 2006–07).

Chapter by One Author in a Book Edited by Another

> 11. William Hallman, "GM Foods in Hindsight," in *Emerging Technologies: From Hindsight to Foresight*, ed. Edna Einsiedel, 13–32 (Vancouver: UBC Press, 2008).

Article in an Encyclopedia

When citing entries in dictionaries and other unsigned, alphabetically arranged reference books, it's best to use *s.v.* (*sub verbo*, "under the word") rather than the volume and page numbers:

> 12. *Encyclopedia of Environment and Society*, 2007 ed., s.v., "Climate".

When an entry is signed, list the author's name first, just as you would for a chapter in an edited book.

Government Document

> 13. Canada. Office of the Auditor General of Canada. *Status Report of the Commissioner of the Environment and Sustainable Development to the House of Commons* (Ottawa: Office of the Auditor General of Canada, 2008), p. 25.

Book Review

> 14. Julie Guthman, review of *Lawn People: How Grasses, Weeds, and Chemicals Make Us Who We Are*, by P. Robbins, *The Professional Geographer*, 60 (2008): 425–426.

Signed Newspaper Article

The place of publication, even if it is not part of the name of the newspaper, should be added in italics along with the title for domestic newspapers, and in parentheses after the title for foreign newspapers:

> 15. David Gow, "Europe to Reaffirm Biofuels Targets," *Guardian* (Manchester), September 10, 2008, A2.

UNSIGNED NEWSPAPER ARTICLE

> 16. "Waterloo Region Bans Bottled Water from Sites," *Toronto Globe and Mail*, September 10, 2008, A5.

In a bibliography entry, the name of the newspaper stands in place of the author.

UNPUBLISHED THESIS OR DISSERTATION

> 17. C. B. Du Preez, "A Mesoscale Investigation of the Seabreeze in the Stellenbosch Winegrowing District" (master's thesis, University of Pretoria, 2007).

A SOON-TO-BE-PUBLISHED WORK

You may be able to see a work that is not yet published, perhaps written by your professor. In this case you will cite the full author, title, and journal/book publisher information. Then, following the anticipated year of publication:

> . . . 2009, in press.

If the year of publication is not yet known, but the work has been accepted for publication:

> . . . forthcoming.

If the work is simply a work in progress, and has not yet been accepted for publication, you might list it as:

> . . . draft copy.

ARTICLE IN AN ONLINE SERIAL

When referencing electronic journals (as in the following example), magazines, or newspapers, follow the format of their print counterparts, with the addition of a URL or DOI. Access dates should be given only in the case of substantive updates, or time-sensitive fields. Include the page range, if it is available, in the bibliography.

> 18. "Sensing in Urban Environments," *Geography Compass* 3 (April 2008), doi: 10.1111/j.1749-8198.2008.0016.x.

WEBSITE

Include as much of the following as can be determined: author of the content; title of the page; title or owner of the site; URL (or DOI); access date, as noted above:

> 19. Chris Severson-Baker, Jennifer Grant and Simon Dyer, "Taking the Wheel: Correcting the Course of Cumulative Environmental Management in the Athabasca Oil Sands," Oil Sands Watch, http://www.oilsandswatch.org/pub/1677 (accessed August 22, 2008).

Notes for subsequent references

1. Subsequent note references should be brief. The preferred style today includes only the author's last name, the title in shortened form, and, if appropriate, the pertinent page number. If you are referring repeatedly to the same source, you can use the abbreviation "Ibid." and the page number (if appropriate), provided that there is no intervening reference to a different source:

> 20. Sayer and Maginnis, "Forests in Landscapes," 109.
>
> 21. D.B. Brooks, in_focus: Water Local-Level Management (IDRC, 2002).
>
> 22. Ibid.
>
> 23. Kevin Bladon et al., "Wildfire Impacts," *Canadian Journal of Forest Research* 38 (2008), 2359–71.
>
> 24. Brooks, *in_focus: Water,* 53.

2. If a bibliography includes all works cited in the document, even the first citation of a particular work can be in shortened form.
3. If you are referring repeatedly to a single primary source, after the first mention (with appropriate identification in a foot- or endnote), you may simply cite the page numbers, enclosed in parentheses, in the text.

Bibliography

A bibliography is an alphabetical list of the works cited in an essay. It is normally an essential part of the note form of documentation. Although a bibliography is not required if you document your references fully in footnotes or endnotes, a full bibliography provides an overview of all the sources cited and can also contain particular relevant works not cited in the essay. The bibliography is always the last item in an essay; any appendices precede it.

FORMAT

The format for bibliographies differs slightly from that for notes:

1. Use a separate page at the end of your essay, with an underlined or italicized heading—*Bibliography*—centred on the page.
2. Single-space all entries, leaving a double space between entries and a triple space between the heading and the first entry. Do not indent entries.
3. Do not number entries; instead, *list them alphabetically* by the author's or editor's surname. If no author is given, begin with the first significant word in the title.
4. Begin each bibliographic entry at the left margin and indent any subsequent line five spaces.
5. To separate the main divisions use periods (rather than the commas and parentheses used in notes).
6. Include all cited works in a single list: books, chapters in books, journal articles, etc.
7. On an annotated bibliography, see p. 71.

ARTICLE IN A JOURNAL

Note that when referencing a work with two or more authors, only the first author's name is inverted:

> Youngblut, Don, and Brian Luckman. "Maximum June-July Temperatures in the Southwest Yukon Over the Last 300 Years Reconstructed from Tree Rings." *Dendrochronologia* 25 (2008): 153–166.

If a work has ten authors or fewer, all should be listed; for references of eleven authors or more, the first seven should be listed, followed by "et al."

BOOK

If you include more than one work by a particular author, list the entries alphabetically by title. Give the name in the first entry only. For subsequent entries, replace the author's name with three dashes followed by a period:

> MacLachlan, Ian. *A Bloody Offal Nuisance: The Persistence of Private Slaughter-Houses in Nineteenth Century London.* Cambridge University Press, 2007.

> ———. *Betting the Farm: Food Safety, Risk Society, and the Canadian Cattle and Beef Commodity Chain.* Toronto: House of Anansi Press, 2004.

Article in an Edited Book

Yamamoto, Daisaku. "Local Tourism Development in a World of Global Interdependence: The Case of Japanese Tourism in Whistler, British Columbia." In *Mountain Resort Planning and Development in an Era of Globalization*, edited by Thomas Clark, Allison Gill and Rudi Hartmann, 41–52. New York: Cognizant, 2006.

Government Document

Canada. British Columbia. Legislative Assembly. *Speech from the Throne: 12 February 2008.*

Unsigned Newspaper Article

If a newspaper article is unsigned, the name of the newspaper stands in place of the author:

Banff Crag and Canyon. "Thompson Lauds Green Banff Biffies," September 9, 2008, 8.

Article in an Online Journal

Northcote, Jeremy. "Spatial Distribution of England's Crop Circles: Using GIS to Investigate a Geo-Spatial Mystery." *Geography Online* 6, no. 2 (2006): 1–15. http://www.siue.edu/GEOGRAPHY/ONLINE/Northcote06.pdf.

Website

Peggie James, President, Soil and Water Conservation Society. "Input to the USDA Climate Change Strategic Plan." Soil and Water Conservation Society. http://www.swcs.org/en/communications/news_releases/input_to_the_usda_climate_change_strategic_plan/.

If an access date is required, it is included parenthetically at the end of the entry.

C. CITATION-NAME

The following examples follow the CSE style. In addition to the parenthetical reference style outlined in section A, the CSE advocates two alternative documentation systems: citation-name and citation-sequence. (See section D for examples of the citation-sequence system of documentation for cartographic, archival, and architectural records.) In the citation-name style, each new book or article mentioned in the text is assigned a number. Then the same number is used for every subsequent reference to the same work. The numbers inserted in the text—preferably as superscript or in parentheses—direct the reader to the reference section at the end of the paper:

> These finding are supported by Becker and Pandle.[4]

> These findings are supported by Becker and Pandle (4).

The first citation of each reference is assigned a number; then at the end of the paper the references are listed in the sequence in which they were cited initially:

IN THE TEXT

> Two authors maintain this stance.[1,2] Lindeman,[3] however, takes particular issue with the suggestion[2] that . . .

IN THE REFERENCE LIST

> 1. Graben A. Greening politics. Luxembourg: Coté; 2006. 233 p.
>
> 2. Singh RL. Rethinking the unthinkable. New Delhi: New Univ Pr; 2004. 341 p.
>
> 3. Lindeman W. 2002. Green politics faces foolishness. Green Bulletin 14(3): 23–46.

The citation-name style has the advantage of interrupting the reading of the text only minimally. The obvious disadvantage is that the reader will have to turn to the reference list to find out which work is being cited.

D. CITATION-SEQUENCE

Maps, architectural records, and various archival sources will include essential information from which you can select what you need for your citations and bibliography:[6]

- author
- title
- edition

- scale
- place of publication
- name of publisher
- date of publication or drawing
- specific material designation and extent (e.g., a map, relief model, or architectural drawing)
- other physical details, such as the material used (e.g., vellum, engraved on copper plate, digital)
- dimensions
- series
- location (if unique, name the specific library or archives, CD-ROM, website)
- insets (if any)
- nature (machine-readable, painting, drawing, etc.)

The following examples are formatted according to the CSE citation-sequence style:

MAP

> *CSE*   Zania. Population over 50 [demographic map]. Zunila: Census Division, Office of International Affairs; 1995. Scale 1:1,500,000. 2 sheets.

MAP IN ATLAS

> *CSE*   Northern Newfoundland. [physical features and thematic sections]. In: The Canadian Atlas: Our Nation, Environment and People. Montreal: The Reader's Digest Association; Ottawa, Canadian Geographic; 2004. p. 124–5. Colour, scale 1:1,000,000.

NOTES

1   Brian Graham, "Ireland and Irishness: Place, Culture and Identity," in *In Search of Ireland: A Cultural Geography*, ed. Brian Graham (London: Routledge, 1997), p. 7.

2   Patrick J. Duffy, "Writing Ireland: Literature and Art in the Representation of Irish Place," in Graham, p. 78.

3   Ellen Churchill Semple, *Influences of Geographic Environment* (New York: Henry Holt, 1911), p. 1.

4   Council of Science Editors, *Scientific Style and Format: The CSE Manual for Authors, Editors and Publishers*, 7th ed. (New York: Cambridge University Press, 2006); American Psychological Association, *Publication Manual of the APA*, 6th ed. (Washington, DC: American Psychological Association, 2010), which is the standard reference for the social, behavioural, and health sciences; and Modern Language Association, *MLA Handbook for Writers of Research Papers*, 7th ed. (New York: The Modern Languages Association of America, 2009). Other specialist style manuals include A.M. Coghill and L.R. Garson, eds., *ACS Style Guide: Effective Communication of Scientific Information*, 3rd ed. (New York: Oxford University Press, 2006); Wallace R. Hansen, ed., *Suggestions to Authors of the Reports of the United States Geological Survey*, 7th ed. (Washington, DC: GPO, 1991), and the American Institute of Physics, *AIP Style Manual*, 4th ed. (New York: American Institute of Physics, 1990). Two standard works are Kate L. Turabian, *A Manual for Writers*, 7th ed. (Chicago: University of Chicago Press, 2007), and *The Chicago Manual of Style*, 15th ed. (Chicago: University of Chicago Press, 2003). The latter is available online.

5   The order of names in a work with several authors may be alphabetical, but in science citations the order generally reflects the authors' relative contributions or their relative standing (e.g., the senior researcher—generally called the principal investigator—first, and his or her assistants thereafter, usually in order of their contributions).

6   Based on Terry Cook, ed., *Archival Citations* (Ottawa: Public Archives of Canada, 1983), p. 5–7.

CHAPTER 12

Writing Examinations

Most students feel nervous before tests and exams. Writing an essay exam—even the open-book or take-home kind—imposes special pressures because both the time and the questions are restricted: you can't write and rewrite the way you can in a regular essay, and you must often write on topics you otherwise would choose to avoid. And, although on the surface objective tests may look easier because you don't have to compose the answers, they force you to be more decisive about your answers than essay exams do. You know that to do your best you need to feel calm—but how? These general guidelines will help you approach any test or exam with confidence. Special advice is provided for open-book and take-home exams, and objective tests.

SPECIAL NEEDS

If, when your instructor identifies the type of examination to be given, you discover you haven't taken that kind of exam before, find out what it will entail and set aside time to practise. This suggestion is not far-fetched: some students have never written an essay exam before they get to college or university, while students from overseas likely never have taken an objective test before coming to North America. If you need help, seek it out—from your instructor or from the student support services—*before* the exam.

If you have need for special consideration when you take an examination, be sure to tell your instructor at the *start* of the term or semester. Don't wait until just before a scheduled exam, since your request may be refused. If you have a documented medical reason to contact the "special needs" or "disabilities" office (the names vary) at your college or university, then do so when you begin your studies or as soon as your condition is identified medically. If necessary, that office can serve as an intermediary between you and your instructor.

BEFORE THE EXAM

Review regularly

Three types of post-lecture reviews were identified in Chapter 3. All three are important. The *immediate post-lecture review* will help to get things organized in your mind; *periodic reviews* will help to get the material and concepts embedded in your memory; and the *full-course review* will sharpen your memory just before the exam. As you review, especially during the periodic and full-course reviews, don't remember just the important material, lecture by lecture, (and all of the readings and any other assignments), but relate all of that material to new information and insights subsequently gained. If you don't review regularly, at the end of the year you'll be faced with relearning rather than remembering.

Set memory triggers

As you review, condense and focus the material by writing down in the margin key words or phrases that will trigger off a whole set of details in your mind. The trigger might be a concept word that names or points to an important theory or definition, or a quantitative phrase such as "three causes of rural poverty" or "five reasons for soil degradation." Sometimes you can create an acronym or a nonsense sentence that will trigger an otherwise hard-to-remember set of facts—something like the acronym HOMES (Huron, Ontario, Michigan, Erie, Superior) for the Great Lakes—though Vermonters, who claim that Lake Champlain is a sixth "Great Lake," might prefer CHOMES! Since the difficulty of memorizing increases with the number of individual items you are trying to remember, any method that will reduce that number will increase your effectiveness.

Ask questions: Try the three-C approach

Think of questions that will get to the heart of the material and cause you to examine the relations between various subjects or issues; then figure out how you would answer them. The three-C approach (pp. 73–5) may be a help. For example, reviewing the *components* of the subject could mean focusing on the main parts of an issue or on the definitions of major terms or theories. When reviewing *change* in the subject, you might ask yourself what elements caused it, directly or indirectly. To review *context* you might consider how certain aspects of the subject—issues, theories, actions, results—compare with others in the course. Essentially, what the three-C approach does is force you to look at the material from various perspectives.

Old examinations are useful both for seeing the types of questions you might be asked and for checking on the thoroughness of your preparation. If old exams aren't available (many professors intentionally keep them out of circulation), you might get together with friends who are taking the same course and ask each other questions. Just remember that the most useful review questions are not the ones that require you to recall facts, but those that force you to analyze, integrate, or evaluate information.

Allow extra time

Be sure you know where the exam will be held, and give yourself lots of time to get there. Nothing is more nerve-wracking than to think that you are going to be late. If you have to travel by car or public transit, don't forget that traffic jams do occur. And alarm clocks can fail to ring (or to wake you up!). Remember Murphy's Law—"Whatever can go wrong will." Anticipate difficulties and allow yourself a good time margin.

Important items

Having extra pens and pencils, an eraser, some tissues, and, if permitted, a ruler, can "save your life" in an exam. You may be permitted to use a calculator and coloured pencils, but check with your instructor beforehand. Be sure to have your student ID card: you may need it in order to be admitted to the exam room or to receive the exam paper.

WRITING AN ESSAY EXAM

Read the exam

An exam is not a hundred-metre dash; instead of starting to write immediately, take time at the beginning to read the instructions and all the questions so that you can devise a plan of action. A few minutes spent on thinking and organizing will produce better results than the same time spent on writing a few more lines.

Apportion your time

Reread the instructions carefully to find out how many questions you must answer and to see whether you have any choice among questions. Subtract five minutes or so for the initial planning, and then divide the remaining time by the number of questions you have to answer. Include in your apportionment the approximate time you will need to draw pertinent maps, diagrams, or graphs—but don't spend too much time on such illustrations unless they are critically important. If possible, allow for a little extra time at the end to reread and edit your work. If the exam instructions indicate that not all questions are of equal value, apportion your time accordingly (work on the most important ones first).

Choose your questions

Decide on the questions that you will answer and the order in which you will do them. Your answers don't have to be in the same order as the questions. If you think you have lots of time, it's a good idea to place your best answer first, your worst answers in the middle, and your second best answer at the end, in order to leave the reader on a high note. If you think you will be rushed, though, it's wiser to work from best to worst. That way you will be sure to get all the marks you can on your good answers, and you won't have to cut a good answer short at the end.

Keep calm

If your first reaction on reading the exam is "I can't do any of it!" force yourself to keep calm; take ten slow, deep breaths as a deliberate relaxation exercise. Decide which question you can answer best. Even if the exam seems disastrous at first, you probably can find one question that looks manageable: that's the one to begin with. It will get you rolling and increase your confidence. By the time you finish, you likely will find that your mind has worked through to the answer for another question.

Read each question carefully

As you turn to each question, read it again carefully and underline all the key words. The wording probably will suggest the number of parts your answer should have; be sure you don't overlook anything (a common mistake when people are nervous). Since the verb used in the question is usually a guide for the approach to take in your answer, it's especially important that you interpret the following terms correctly:

- **explain:** show the *hows* or *whys*;
- **compare:** give both similarities and differences—even if the question doesn't say "compare and contrast";
- **outline:** state simply, without much development of each point (unless specifically asked); and
- **discuss:** develop your points in an orderly way, taking into account contrary evidence or ideas.

Make notes

Before you even begin to organize your answer, jot down key ideas and information related to the topic on rough paper or the unlined pages of your answer book. These notes will spare you the worry of forgetting something by the time you begin writing. Next, arrange the points you want to use into a brief plan, using numbers to indicate their order (that way, if you change your mind, it will be easy to reorder them). At the close of the exam be sure to cross out these notes so that the evaluator won't think they are your actual answers. You will have to submit any notes with the rest of your paper.

Be direct

Get to the points quickly and illustrate them frequently. In an exam, as opposed to a term paper, it's best to use a direct approach. Don't worry about composing a graceful introduction: simply state the main points that you are going to discuss, and then get on with developing them. Remember that your paper is just one of many read and marked by someone who has to work quickly—the clearer your answers are, the better they will be received. For each main point give the kind of specific details that will prove you really know the material. General statements will show you are able to assimilate information, but they need examples to back them up.

Write legibly

Bad handwriting makes readers cranky. When the marker has to struggle to decipher your ideas, you may get poorer results than you deserve. If for some special reason (such as a physical handicap) your writing is hard to read, ask whether you can make special arrangements to use a computer or, in special cases, a tape recorder. If your writing is just plain bad, it's probably better to print. Although some instructors will allow pencils, others will demand you write in ink.

Write on alternate lines

Writing on every other line will make your writing easier to read, and will leave you space for changes and additions; you won't have to cover your paper with a lot of messy circles and arrows.

Keep to your time plan

Keep to your plan and don't skip any questions. Try to write something on each topic. Remember that it's easier to score half marks for a question you don't know much about than it is to score full marks for one you could write pages on. If you find yourself running out of time on an answer and still haven't finished, summarize the remaining points and go on to the next question. Leave a large space between questions so that you can go back and add more if you have time. If you write a new ending, remember to cross out the old one—neatly.

Reread your answers

No matter how tired or fed up you are, reread your answers at the end if there is time. Check especially for clarity of expression; try to get rid of confusing sentences and increase the logical connection between your ideas. Revisions that make answers easier to read are always worth the effort.

Check question numbers

During your final check, ensure that each of your answers starts with the number that correctly links it to the question you have addressed. An unnumbered answer may not be graded.

WRITING AN OPEN-BOOK EXAM

If you think that permission to take your books into the exam room will guarantee you success, be forewarned: you could fall into the trap of relying too heavily on them. You may spend so much time rifling through pages and looking things up that you won't have time to write good answers. The result may be worse than if you had been allowed no books at all.

If you want to do well, use your books only to check information and look up specific, hard-to-remember details for a topic you already know a good deal about. For instance, if the exam is on fluvial processes, you could check on detailed statistical formulae; if, in a political geography exam, the issue is electoral studies, you may want to look up voting statistics; for an exam in social theory you could check some classical references and find the authors' exact definitions of key concepts—if you know where to find them quickly. In other words, use the books to make sure your answers are precise and well illustrated. Never use them to replace studying and careful exam preparation.

WRITING A TAKE-HOME EXAM

The time you are given for a take-home exam will be limited to, say, one afternoon, or, perhaps, overnight. Whatever it is, plan accordingly and carefully: time for reviewing the question or questions; time for culling from your notes and pertinent source material; time for organizing your thoughts; time for writing; time for reviewing, rewriting, and making any needed corrections; and time for submitting your work before the deadline. If the time is short, you will just have to charge ahead, but do leave time for rereading what you have written, so you can catch any errors. If the period you are given is overnight, or even longer, remember to build in time for eating, sleeping, and other breaks. No matter what the length of time you are given to write the exam, it seems like it's never "enough"! Try to avoid working yourself into a frenzy. Plan, and proceed according to the plan, always with an eye on the clock.

As with other exams, you will be expected to demonstrate your command of the material. Don't go looking for fresh sources if time is very limited. Since you will have been doing your periodic reviews on a regular basis, you will have a good idea of where to look in the sources you have at hand. Use that knowledge to get the best result. Keep in mind that a take-home exam is intended to test your overall command of the course's concepts, methods, and subject matter. Your reader is less likely to be concerned with your specialized research than with evidence that you have understood and assimilated what you have learned.

The guidelines for a take-home exam are similar to those for a regular exam; the only difference is that you don't need to keep *such* a close eye on the clock. Be sure to:

1. Keep your introduction to each answer short and get to the point quickly.
2. Organize your answer in a straightforward and obvious way, so that the reader easily sees your main ideas.
3. Use frequent concrete examples to back up your points.
4. Where possible, show the range of your knowledge of the course material by referring to a variety of sources, rather than constantly using the same ones.
5. Try to show that you can analyze and evaluate material, that you can do more than simply repeat information.
6. If you are asked to acknowledge the sources of any quotations you use, be sure to jot them down as you go; you may not have time to do so at the end of the exam.
7. Remember that you may be asked to provide a list of all the works you consulted while writing the exam. If so, attach a list of them to your exam.

WRITING AN OBJECTIVE TEST

Objective tests are common in the sciences and social sciences. Although sometimes the questions are the true-false kind, most often they are multiple-choice. The main difficulty with these tests is that their questions are designed to confuse students who aren't certain of the correct answers. If you are forever second-guessing yourself or you readily see two sides to every question, you may find objective tests particularly hard at first. Fortunately, practice almost always improves performance.

Preparation for objective tests is the same as for other kinds. Here, though, it's especially important to pay attention to definitions and unexpected or confusing pieces of information, because they can so readily be adapted to make objective-test questions. Although there is no sure recipe for doing well on an objective test—other than knowing the course material completely and confidently—these suggestions may help you to do better:

Find out the marking system

If marks are based solely on the number of correct answers, you should pick an answer for every question even if you aren't sure it's the right one. For a true-false question you have a 50 per cent chance that it will be correct; and even for a multiple-choice question with four possible answers, you would get an average of 25 per cent right if you picked the answers blindfolded.

On the other hand, if there is a penalty for wrong answers—if marks are deducted for errors—you should guess only when you are fairly sure you are right, or when you are able to rule out most of the possibilities. Don't make wild guesses.

Do the easy questions first

Go through the test at least twice. On the first round, don't waste time on troublesome questions. Since the questions usually are of equal value, it's best to get all the marks you can on the ones you find easy. You can tackle the more difficult questions on the next round. This approach has two advantages:

1. You won't be forced, because you have run out of time, to leave out any questions that you easily could have answered correctly.
2. When you come back to a difficult question on the second round, you may find that you have figured out the answer in the meantime.

Make your guesses educated ones

If you decide to guess, at least increase your chance of getting the answers right. Forget about intuition, hunches, and lucky numbers. More important, forget about so-called patterns of correct answers—the idea that if there have been two "A" answers in a row, the next one can't possibly be "A," or that if there hasn't been a "true" for a while, "true" must be a good guess. Unfortunately, many test-setters either don't worry about patterns at all, or else deliberately mislead pattern-hunters by giving the right answer the same letter or number several times in a row.

Remember that constructing good objective tests is a special skill that not all instructors have mastered. Often the questions they pose, though sound enough as questions, do not produce enough realistic alternatives for answers. In such cases the test-setter may resort to some less-realistic options, and if you keep your eyes open you can spot them. James F. Shepherd has suggested a number of tips that will increase your chances of making the right guess— that is, if you are quite sure you don't know the correct answer:[1]

- Start by weeding out all the answers you know are wrong, rather than looking for the right one.
- Avoid any terms you don't recognize. Some students are taken in by anything that looks like sophisticated terminology and may assume that such answers must be correct. They are usually wrong: the unfamiliar term may well be a red herring, especially if it is close in sound to the correct one.
- Avoid extremes. Most often—but not always—the right answer lies in between. For example, suppose that the options are the numbers 800,000; 350,000; 275,000; and 15: the highest and lowest numbers are likely to be wrong.
- Avoid absolutes, especially on questions dealing with people. Few aspects of human life are as certain as is implied by such words as *everyone*, *all*, or *no one*; *always*, *invariably*, or *never*. Statements containing these words usually are false.
- Avoid jokes or humorous statements.
- Avoid demeaning or insulting statements. Like jokes, these are usually inserted simply to provide a full complement of options.
- Choose the best available answer, even if it is not indisputably true.
- Choose the long answer over the short (it's more likely to contain the detail needed to make it right) and the particular statement over the general (generalizations are usually too sweeping to be true).
- Choose "all of the above" over individual answers. Test-setters know that students with a patchy knowledge of the course material will often fasten on the one fact they know. Only those with a thorough knowledge will recognize that all the answers listed *may* be correct.

SOME FINAL TIPS

Remember to keep your eye on the clock so you always know how much time you have remaining. If you have time at the end of the exam, go back and reread the questions. One or two wrong answers caused by misreading can make a significant difference to your score. On the other hand, don't start second-guessing yourself and changing a lot of answers at the last minute. Studies have shown that when students make changes they are often wrong (your first answer often is the correct one). Stick to your decisions unless you know for certain you have made a mistake.

NOTE

1   James F. Shepherd, *College Study Skills*, 6th ed. (Boston: Houghton Mifflin, 1997) and, idem., *RSVP: The College Reading, Study, and Vocabulary Program*, 5th ed. (Boston: Houghton Mifflin, 1995).

 CHAPTER 13

Words: Gender, Race, and Other Sensitivities

Words convey meaning. The words you use give shape to the world in which you live. But are you sure that the words you use convey the meanings you intend, with due consideration for the way they will be read and interpreted by others? Certain words may mean one thing to one reader and quite a different thing to another. To "rap," for example, to someone of your grandparents' generation, may mean to make a smart, sharp blow with the hand on a hard surface; to someone who grew up in the 1960s, it may mean to talk, while to someone younger it likely means to sing a rapid rhyming song. The meanings of some words change slowly, others quickly, and new words are forever coming into popular use, so dictionaries have to be updated periodically in new editions. Glance through the *Oxford English Dictionary*, the *Shorter Oxford English Dictionary on Historical Principles*, or any other sound dictionary that gives etymological information and you will see the roots of words, when they were first used, what they meant then, and how they may have changed in meaning through time. We live in an age of heightened sensitivity to people's rights, and in which people readily take offence when they feel slighted. It's essential to be aware of how the words you use may be interpreted. This chapter focuses on some issues related to words and the importance of choosing them carefully.

CULTURAL CONVENTIONS

Cultural conventions guide us in the ways and the situations in which we may use certain words. Slang, swearing, and other expressions used "on the street" are usually unacceptable in other settings, such as the classroom or a formal essay. A minute's reflection will bring to mind many words that may offend other people, especially in formal situations. Be aware, too, that using certain words may lead to misunderstanding, at a minimum, and at worst, some words may convey bigotry and intolerance.

Language is powerful. It can reinforce and pass on attitudes, good or bad; derogatory language, for instance, can reinforce discriminatory attitudes, actions, and institutions. Paying attention to the words we use can help us to examine our own prejudices, attitudes, and values, and even our actions toward others. As a researcher, it is essential that you understand your own biases and thought patterns. Knowing the importance of words as conveyers of meaning will help you achieve clarity of expression in a manner that is not offensive to others. In so doing, you may find that it is necessary to challenge conventional usage. In short, don't assume that "cultural conventions" are always "correct."

"MAN" AND . . .

Until the 1970s there were many courses and books with titles such "Man and the Land," or "Man and Environment." "Man" was taken to mean *Homo sapiens* in general, all of humankind. Yet in much of the older literature it is clear that women were indeed neglected, in research and in public life generally. While in some cases the absence of reference to women may be an accurate reflection of the author's intent, in others the use of "man" is totally misleading. For instance, to say that "ancient man developed agriculture" is to ignore the fact that in hunting and gathering societies the females largely were responsible for the domestication of plants. Hence, in this instance, it would be more accurate to use the term ancient people. Another example comes from an early twentieth-century geography book: ". . . man carries on his devastating activity particularly in forested regions." What does "man" mean in this sentence? Men? Humans? All human societies, or only some? The same book, in a discussion of materials used for houses, states that "man finds them equally suitable for his work and for his needs." Here again, you must use your imagination to realize that the author was using "man" in the generic sense, to mean human beings in general.

Today, however, it is acknowledged that the use of "man" is inappropriate when the reference is to people of both sexes: use humankind, or people, or women and men instead. Try to avoid the use of compound words starting or ending with "man." For instance, you can use police officer for "policeman," firefighter for "fireman," synthetic or artificial for "man-made," English people for "Englishmen," and so on, if the term is intended to be inclusive. If gender is pertinent, on the other hand, specific terms such as policewomen and policemen are appropriate: don't hide gender identity when it is important for your reader to know it. (Be aware, too, that some words that may appear to be based on "man," such as "manage," "manipulate," "manufacture," "manner," and "manual," are in fact derived from the Latin *manus*, "hand," so there is no need to avoid them.)

Gender-inclusive terms are easy to find, as with the average person for "the average man," and workforce or personnel for "manpower" or "workmen." Where pronouns are concerned, simply changing a singular to a plural often takes care of the problem (see Pronoun agreement and gender, pp. 237–8). Excessive dedication to gender-inclusive language can lead to awkward terms such as "horsepersonship". Though occasionally you may be able to use humour to advantage, in most cases you can express the same idea in different words (e.g., in this case, "equestrian skill").

Sometimes avoidance of a certain word can cause a loss in meaning; for instance, a "manhole cover" is generally found in a street, whereas a "maintenance cover" could be anywhere. In such cases you may have to add extra information in order to make your meaning clear. And, of course, sometimes it is necessary to use gender qualifiers, as in "The number of female geographers is increasing."

In addition to avoiding use of sexist language, you must be careful not to demean others through your choice of words. For example, don't trivialize women by making foolish ("The male TA is better than the female TA because males are better at science") or stereotypical ("Women are caring, men are not") generalizations. Pay attention to what you are writing and be aware of the assumptions underlying your choice of words.

Some students confuse "sex" with "gender." "Sex," as a noun, refers to either of the two main divisions (male and female) into which living things are placed on the basis of their reproductive functions; as an adjective, "sex" refers to sexual characteristics or instincts (such as sexual activity, sex appeal, and so on). "Gender," as a noun, refers either to the grammatical classification of words (masculine, feminine, or neuter, as in "it") or, notably, to psychological associations, or other characteristics (for example, role differentiation) that reveal behavioural dimensions of being. Be careful to select the appropriate word.

RACE AND CULTURE, AND RACIST LANGUAGE

Two words too often used incorrectly as synonyms are "race" and "culture." In the nineteenth century, "race" was almost synonymous with what today is called "culture" (a German term, coined only in the late 1800s, that did not come into the English language until the late 1920s). Today, however, "race" refers strictly to a (human or animal) population that is distinguished as a more or less distinct group by genetically transmitted characteristics and therefore deemed to be biologically and physically identifiable. In the past, people were identified as belonging to racial categories (Caucasian, Mongolian, Negroid), but these categories increasingly are challenged as new understandings are reached about human genetics. Indeed, DNA testing has revealed that such categories of "race" are grossly limited or false, because all populations are, to varying degrees, intertwined. Racial categories, as used in some censuses, for example, are at best socio-political categories, not scientific ones.

"Culture" refers to the learned attributes that make one group of people distinct from another. These attributes include features such as language, religion, rituals, customs, celebrations, and ways of doing things. For example, to refer to French Canadians as a particular "race" is incorrect, since they are not racially distinct from the majority of other Canadians (most are Caucasians), although they do have some cultural differences. And, of course, many French Canadians refer incorrectly to most others in Canadian society as "English," irrespective of actual national origins. Be precise: use "race" when it is the appropriate term, and "culture" likewise. Don't mix them up.

Use of language that is racist may imply—or be read by others as implying—that you accept discriminatory attitudes toward others based on race. Don't use descriptors when they aren't required (e.g., "two black men" or "two Asian women" when "two men" or "two women" is adequate) to convey pertinent information in certain situations. In general, it is inappropriate to use derogatory words or racial slurs and adjectives to describe people.

ETHNICITY AND "VISIBLE MINORITIES"

Associated with a concern for race and culture is the need to be aware of the appropriate words to use when referring to people with attributes that differentiate them from others in society. For example, in Canada many people speak English or French with an accent, observe holidays or traditions that differ from local norms, or are members of "visible minorities." The term "South Asian," for instance, is used to refer in a general way to people from several different countries, including India, Pakistan, Bangladesh, and Sri Lanka. Even within each of these countries, significant differences of language, religious beliefs and practices, dress, and customs exist. At minimum, it is important to be aware of a person's country of origin, and to know that a Sri Lankan will not want to be called an Indian, or vice versa. The same holds true for other parts of the world. Japanese and Chinese people, for instance, are different from one another, and would not appreciate being identified incorrectly. Belgians are not French—not even when they are Walloons (as opposed to Flemings). And while Botswana is in southern Africa, the people of that country are emphatically not South Africans. At times, of course, it may be correct to link peoples—for example, from southern Africa, or northern Europe, or South Asia—if the discussion pertains to a whole region and not simply to part of it. But keep in mind the danger of cultural "blindness": there is a tendency for North Americans to generalize about Third World countries—for instance, to consider a reference specific to Zambia as being representative of all Africa—in a way they would never do where a European country (say, Scotland or Germany) is concerned. Also, beware of generalizing about people's belief systems: not all Arabs are Muslims, and not all Muslims are Arabs; not all Jews are Zionists; not all socialists are communists; and so on.

What about "Indians" in Canada and the United States? Particularly in governmental and bureaucratic situations, the word Indian is used to refer to people indigenous to North America (as opposed to those who have come, or whose ancestors came, from India in South Asia). The terms Native people (in Canada), Native Americans (in the U.S.A.), Aboriginal people, and Indigenous people also are used. The term "people" has international legal meaning; hence Canada's "First Nations" intentionally use the term people to describe themselves. Likewise, whereas Europeans used to refer to the Native people of the Arctic as Eskimos, the people themselves prefer the term Inuit, which means simply "the people." And, depending upon the context, it may be more appropriate to use the specific name—Cree, Naskapi, Nisga'a, Anishnabe, Oneida, Chippewa, Navajo, etc.—than a broad term such as Indigenous people. In Canada as well as the U.S.A., some Indigenous people refer to themselves simply as Americans—but do not consider themselves to belong to either country. In short, be aware of what people call themselves, and use the appropriate term. Relatedly, do not use slurs and adjectives that are derogatory and demeaning—and could lead you to be seen as a racist, even if you believe you are not.

"WHITES" AND "BLACKS" AND . . .

Aboriginal people in Canada often use the term "white" to refer to anyone who is not Aboriginal—including "non-whites." "White" often is used by "whites" to refer to themselves—to the exclusion of all "non-whites"—although contemporary governments generally do not use or accept either term. Unlike non-Maori New Zealanders, who have adopted the Maori word *pakeha* for people of European origin, non-Aboriginal Canadians have not yet found a suitable term to describe themselves. Similar problems arise in designating the people in both Canada and the United States who formerly were referred to as "Negro" or "coloured." These terms reflected the enslavement of their ancestors but were not self-chosen; instead they were applied to the population by "whites." As a new, revitalized sense of individual and group self-worth developed in the 1960s, a determined effort was made by the people themselves to find an acceptable descriptor. First Afro-American and then African American—and, similarly, Afro-Canadian and African Canadian—became popular, but Black has also come into widespread usage (though you should be aware that not everyone today accepts Black either; some prefer people of colour). In Britain, however, Black is used inclusively to refer to people of African, Caribbean, and South Asian origin; as you read reference sources, and as you write, keep this distinction in mind. Above all, once again, use the words that the people in question prefer, and avoid prejudicial, discriminatory terms.

DISABILITIES

The advocacy movement for people with disabilities (not "the disabled") continues, yet many members of the general public remain unsure of the preferred terms. Here is a guide to some of them:

Do not use	Do use
afflicted	person with a disability; visually impaired; etc.
confined to a wheelchair	person who uses a wheelchair; wheelchair-user.
cripple	person with a mobility impairment, a spinal-cord injury, arthritis, etc.
differently able	person with a disability.
handicapped	person with a disability (unless referring to an environmental or attitudinal barrier, in which case "handicapped" is appropriate).

normal	person who is non-disabled ("normal" is acceptable only in reference to statistics, e.g., "the norm").
suffers from . . .	use the appropriate verb: *has* cerebral palsy, *is* hard of hearing, etc. (Disability is not synonymous with suffering.)

What is the common denominator in the right-hand column? In each of these examples the emphasis is on the person, not the disability. Remember that disabilities are incidental; they do not define people.

Some terms, such as "deaf," are in common use but it might sometimes be appropriate to use the phrases "person with a hearing impairment" or "person who is hard of hearing," if the disability is not total. The word "blind" is also used, but there are some people who are "visually impaired" as opposed to being totally blind. Know who it is you are referring to, and the nature of the disability they live with.

CONCLUSION

It's essential to be attentive to the words you use when describing people. This chapter has identified a number of terms that may be hurtful, offensive, and divisive (and that could cause outright rejection of your work). But the list here is far from complete. How can you avoid other discriminatory terms? Although there is no foolproof method, in practically every case the best advice is simply to find out what term the people you are writing about prefer: the more you know about any subject—including people—the less likely you are to make mistakes. An even better solution is, when possible, to do research *with* the people in question, rather than *about* or "*on*" them.

CHAPTER 14

Writing with Style

Writing with style does not mean stuffing your prose with fancy words and extravagant images. Any style, from the simplest to the most elaborate, can be effective, depending on the occasion and intent. Writers known for their style are those who have projected something of their own personality into their writing; we can hear a distinctive voice in what they say. Obviously it takes time to develop a unique style. To begin, you have to decide what general effect you want to create. Journalists have led the trend toward short, easy-to-grasp sentences and paragraphs. Writing in an academic context, by contrast, you may expect your audience to be more reflective than the average newspaper reader, but the most effective style is still one that is clear, concise, and forceful.

BE CLEAR

Since sentence structure is dealt with in Chapter 15, this section focuses on the use of clear words and the creation of clear paragraphs.

Choose clear words

The key to good writing is the use of clear (easy to understand, unambiguous) and appropriate words. Think carefully about what you want to say and then write it out in words that convey your meaning accurately and effectively. Remember to refer to specialized disciplinary dictionaries. For hints on using dictionaries and thesauruses built into computer programs, see Chapter 1. On the changing meanings of certain words related to gender, race, and so on, see Chapter 13.

Use plain English

Frequently, plain words are more forceful than fancy ones. If you aren't sure what plain English is, think of everyday speech (apart from profanities): how do you talk to your friends? Many of our most common words—the ones that sound most natural and direct—are short. A number of them are among the oldest words in the English language. By contrast, most of the words that

English has derived from other languages are longer and more complicated; even after they've been used for centuries, some can still sound artificial. For this reason you should be aware of words loaded with prefixes (*pre-*, *post-*, *anti-*, *pro-*, *sub-*, *maxi-*, etc.) and suffixes (*-ate*, *-ize*, *-tion*, etc.). These Latinate attachments can make individual words more precise and efficient, but putting a lot of them together will make your writing seem dense and hard to understand. In many cases you can substitute a plain word for a fancy one:

Table 14.1 Examples of "fancy" and "plain" English

Fancy	Plain	Fancy	Plain
accomplish	do	oration	speech
cognizant	aware	prioritize	rank
commence	begin, start	proceed	go
conclusion	end	remuneration	pay
determinant	cause	requisite	needed
fabricate	build	sanitize	clean
finalize	finish, complete	subsequently	later
maximization	increase	systematize	order
modification	change	terminate	end
numerous	many	transpire	happen
obviate	prevent	utilize, utilization	use, use

Suggesting that you write in plain English does not mean that you should never pick an unfamiliar, long, or foreign word: sometimes those words are the only ones that will convey precisely what you mean. Inserting an unusual expression into a passage of plain writing may be an effective way to catch the reader's attention—as long as you don't do it too often. And, of course, writing clearly does not mean that you automatically should avoid all words that have specific scientific meanings. In many circumstances, such words are invaluable for conveying precise and appropriate meanings, and using them correctly will indicate your understanding. Just remember that when you use technical language your instructors will not be impressed by the mere presence of such words; appropriate disciplinary terminology must be used correctly (see "Jargon" below). In other contexts—for example, if you are reworking an essay for submission to a newspaper—it is wise to avoid scientific terms that may not be familiar to the general reader.

Be precise

Always be as precise or exact as you can. Avoid all-purpose adjectives such as *major*, *significant*, and *important*, and vague verbs such as *involve*, *entail*, and *exist*, when you can be more specific. The two pairs of sentences, following, illustrate how precision may be improved:

> *orig.* Wagner and Mikesell were involved in producing the influential book *Readings in Cultural Geography.*
>
> *rev.* Wagner and Mikesell co-edited the influential book *Readings in Cultural Geography.*
>
> *orig.* Burning petrochemicals has a significant impact on the environment.
>
> *rev.* Burning petrochemicals has a detrimental impact on the environment.

AVOID UNNECESSARY QUALIFIERS

Qualifiers such as *very*, *rather*, and *extremely* are overused. Experienced writers know that saying something is *very beautiful* may have less impact than saying simply that it is *beautiful*. For example, compare these sentences:

> They devised an elegant hypothesis to explain the data.
>
> They devised a very elegant hypothesis to explain the data.

Which has more punch? Here are a few more examples of unnecessary qualifiers:

hard evidence	evidence
mountainous in character	mountainous
positively rejected	rejected
really dangerous	dangerous
truly substantial	substantial

When you think that an adjective needs qualifying—and sometimes it will—first see if it is possible to change either the adjective or the phrasing, and then try to write a precise statement. Instead of writing:

> Maple syrup production was really great last year,

write a statement such as:

> Maple syrup production last year was 30 per cent higher than the year before.

In some cases qualifiers not only weaken your writing but are redundant, since the adjectives themselves are absolutes. To say that something is *very unique* makes as much—or as little—sense as to say that someone is *rather pregnant* or *very dead*.

CUT REDUNDANT WORDS

Avoid saying the same thing twice (redundant repetition is known as tautology). For example:

Tautology	Meaning
each individual person	each person
enclosed with this essay	enclosed
grouped together	grouped
I personally agree	I agree
my own personal opinion	my opinion
percolated down	percolated
postponed until another time	postponed
the reason for this is because	because

JARGON

All academic subjects have their own terminology or jargon: special, technical language. It may be unfamiliar to "outsiders," but it helps specialists explain things to each other. Precise disciplinary jargon is appropriate for informed audiences, such as your professors and the students in your lab or seminar section. Remember, though, that you must be certain of what a specialized term means, and that you shouldn't use jargon unnecessarily, simply to make yourself seem more knowledgeable. Too often the result will be not clarity but complication. The guideline is easy: use specialized terminology only when it's a kind of shorthand that will help you explain something more precisely and efficiently to a knowledgeable audience. If plain prose will do just as well, stick to it, especially when your audience is unlikely to know the disciplinary jargon.

Create clear paragraphs

Paragraphs come in so many sizes and patterns that no single formula could possibly cover them all. There are three basic principles to remember:

1. A paragraph is a means of developing and framing an idea or impression.
2 The divisions between paragraphs are not random, but indicate a shift in focus.
3. Paragraphs normally include at least two sentences.

With these principles in mind, you should aim to include three elements in each paragraph:

- the topic sentence, to tell the reader the subject of the paragraph;
- a supporting sentence or sentences, to convey evidence or develop the argument; and
- a conclusion, to indicate to the reader that the paragraph is complete. The conclusion may summarize the paragraph, especially if the discussion is long or complex, and may include a reference (direct or indirect) to the topic sentence.

Keep these points in mind as you write. The following sections offer additional advice.

DEVELOP YOUR IDEAS

You are not likely to sit down and consciously ask yourself, "What pattern shall I use to develop this paragraph?" What comes first is the idea you intend to develop: the pattern the paragraph takes should flow from the idea itself and the way you want to discuss or expand it. (The most common ways of developing an idea are outlined on pp. 79–82.)

You may take one or several paragraphs to develop an idea fully. For a definition alone you could write one paragraph or ten, depending on the complexity of the subject and the nature of the assignment. Just remember that ideas need development, and that each new paragraph signals a change in idea.

THE TOPIC SENTENCE

Skilled skim-readers know that they can get the general idea of a book simply by reading the first sentence of each paragraph. The reason is that most paragraphs begin by stating the central idea to be developed. If you are writing your essay from a formal plan, you probably will find that each section and subsection will generate the topic sentence for a new paragraph.

Like the thesis statement for the essay as a whole, the topic sentence is not obligatory: in some paragraphs the controlling idea is not stated until the middle or even the end, and in others it's not stated at all but merely implied. Nevertheless, it's a good idea to think out a topic sentence for every paragraph. That way you'll be sure that each one has a readily graspable point and is connected clearly to what comes before and after. When revising your initial draft, check to see that each paragraph is held together by a topic sentence, either stated or implied. If you find that you can't formulate one, you probably should rework the whole paragraph.

FOCUS

A clear paragraph will help to focus attention and to facilitate understanding for yourself and the reader. The lead sentence normally identifies the central idea of the paragraph; and, to maintain focus, what follows should contain only details that are linked to the central idea. In addition, each paragraph should be structured so readers can see that the details are related. A good way to maintain focus is to keep the same grammatical subject in most of the sentences that make up the paragraph. When the grammatical subject is shifting all the time, a paragraph loses focus, as in the following example:

> **orig.** In metropolitan centres, <u>main street redevelopment</u> in favour of major shopping-hotel-convention centres or large cultural complexes has become almost obligatory. <u>Rehabilitation</u> of the oldest and often the most distinctive parts of the city centre, typically identifiable with the "zone of discard" of the central business district, often takes place at the same time. <u>Economic, demographic, and cultural factors</u> of international relevance have provided the impetus for such changes. <u>Energy concerns and the return of the urban lifestyle</u> as a fashionable ideal among an influential elite are some of the factors. <u>Substantial dislocation</u> of the physical, functional, and social components of the inner city takes place, mostly at the expense of its poorer residents, as redevelopment occurs.

Most readers would lose their focus on the central message of this paragraph because the grammatical subject (underlined) constantly jumps from one thing to another. Notice how much stronger the focus becomes when all the sentences have the same grammatical subject—either the same term, a synonym, or a related pronoun:

rev. Main street redevelopment in metropolitan centres now favours major shopping-hotel-convention centres or cultural complexes. Typically such redevelopment takes place as the oldest parts of the city centre (typically the "zone of discard" of the central business district) are rehabilitated. The revitalization of inner cities is a response to economic, demographic, and cultural factors of international relevance, and reflects changing energy concerns and the return of the urban lifestyle as a fashionable ideal for the influential elite. Whatever its merits, the revitalization process entails a substantial dislocation of the physical, functional, and social components of the inner city, mostly at the expense of its poorer residents.

Naturally it's not always possible to retain the same grammatical subject (focus) throughout a paragraph. If you were comparing the athletic pursuits of boys and girls, for example, you would have to switch back and forth between boys and girls as your grammatical subject. In the same way, your focus would need to shift if you were discussing examples of an idea or exceptions to it.

Avoid Monotony

If most or all of the sentences in your paragraph have the same grammatical subject, how do you avoid boring your reader? There are two easy ways:

- **Use stand-in words.** Pronouns, either personal (*I, we, you, he, she, it, they*) or demonstrative (*this, that, those*), can stand in for the subject, as can synonyms (words or phrases that mean the same thing). Stand-in words can be useful, but be aware that pronouns may reduce the clarity of your writing (because pronouns are less precise than the grammatical subject).
- **"Bury" the subject by putting something in front of it.** When the subject is placed in the middle of the sentence rather than at the beginning, it is less obvious to the reader. If you take another look at the revised paragraph above, you'll see that in some sentences there is a word or phrase in front of the subject, giving the paragraph

a feeling of variety. Even a single word, such as *first, then, lately,* or *moreover*, will do the trick. (Incidentally, this is a useful technique to remember when you are writing a letter of application and want to avoid starting every sentence with *I*.)

LINK YOUR IDEAS

To create coherent paragraphs, you need to link your ideas clearly. Linking words are those connectors—conjunctions and conjunctive adverbs—that show the *relations* between one sentence, or part of a sentence, and another; they also are known as transition words, because they bridge the transition from one thought to another. Make a habit of using linking words when you shift from one grammatical subject or idea to another, whether the shift occurs within a single paragraph or as you move from one paragraph to the next. See Table 14.2 for examples of commonly used connectors and the logical relations they indicate. Note that numerical terms such as *first, second,* and *third* also work well as links.

VARY THE LENGTH, BUT AVOID EXTREMES

Ideally, academic writing will have a comfortable balance of long and short paragraphs. Avoid the extremes—especially the one-sentence paragraph, which can state only an idea, without explaining or developing it. A series of very short paragraphs is usually a sign that you have not developed your ideas in enough detail, or that you have started new paragraphs unnecessarily. On the other hand, a succession of long paragraphs can be tiring and difficult to read. In deciding when to start a new paragraph, remember always to consider what is clearest and most helpful for the reader.

BE CONCISE

At one time or another, you may be tempted to pad your writing. Whatever the reason—because you need to write two or three thousand words and have only enough to say for one thousand, or just because you think length is strength and hope to get a better mark for the extra words—padding is a mistake. You may fool some of the people some of the time, but you are not likely to impress a first-rate mind with second-rate verbiage.

Strong writing always is concise. It leaves out anything that does not serve some communicative or stylistic purpose, in order to say as much as possible in as few words as possible. Concise writing will help you do better on both your essays and your exams.

Table 14.2 Examples of connectors and their logical relations

Connectors (Linking Words)	Logical Relation
again also and furthermore in addition likewise more moreover similarly	addition to previous idea
although but by contrast despite even so however in spite of nevertheless on the other hand rather yet	change from previous idea
accordingly consequently for this reason hence so thus	summary or conclusion

Guidelines for concise writing

USE ADVERBS AND ADJECTIVES SPARINGLY

Avoid the shotgun approach to adverbs and adjectives: don't include numerous modifiers unless you are sure that they clarify meaning. One well-chosen word is always better than a series of synonyms:

orig. As well as being <u>costly</u> and <u>financially</u> <u>extravagant</u>, the venture is <u>reckless</u> and <u>foolhardy</u>.

rev. The venture is <u>foolhardy</u> as well as <u>costly</u>.

AVOID NOUN CLUSTERS

A trend in some writing is to use nouns as adjectives, as in the phrase *noun cluster*. This device occasionally can be effective, but frequent use can produce a monstrous pile-up of words. Breaking up noun clusters may not always produce fewer words, but it may make your writing easier to read:

orig.   word processor utilization manual

rev.   manual for using word processors

orig.   pollution investigation committee

rev.   committee to investigate pollution

AVOID CHAINS OF RELATIVE CLAUSES

Sentences full of clauses beginning with *which*, *that*, or *who* typically are more wordy than necessary. Try reducing some of those clauses to phrases or single words:

orig.   The solutions <u>which</u> were discussed last night have a practical benefit <u>which</u> is easily grasped by people <u>who</u> have no technical training.

rev.   The solutions discussed last night have a practical benefit, easily grasped by non-technical people.

TRY REDUCING CLAUSES TO PHRASES OR WORDS

Independent clauses can be reduced by subordination. Here are a few examples:

orig.   The report was written in a clear and concise manner and it was widely read.

rev.   Written in a clear and concise manner, the report was widely read.

rev.   Clear and concise, the report was widely read.

orig.   His plan was of a radical nature and was a source of embarrassment to his employer.

rev.   His radical plan embarrassed his employer.

For more detail on subordination and reduction, see pp. 226–7.

STRIKE OUT HACKNEYED EXPRESSIONS AND CIRCUMLOCUTIONS

Trite or roundabout phrases may flow from your pen without a thought, but they make for stale prose. Unnecessary words are dead wood; be prepared to hunt and chop ruthlessly to keep your writing vital:

Wordy	Revised
at this point in time	now
consensus of opinion	consensus
due to the fact that	because
examined on a day-to-day basis	examined daily
if at all possible	if possible
in all likelihood	likely
in the eventuality that	if
in the near future	soon
it goes without saying	obviously
there was a large degree of agreement	most people agreed
we were in the process of testing	we were testing
when all is said and done	(omit)

AVOID BEGINNING WITH "IT IS" AND "THERE IS"

Although it is not possible always, try to avoid beginning your sentences with *it is* or *there is (are)*. Your sentences will be crisper and more concise if you do:

orig. There is little time remaining for the industrialized countries of the world to reverse the effects of carbon dioxide on the ozone layer.

rev. Little time remains for the industrialized countries to reverse the effects of carbon dioxide on the ozone layer.

orig. It is certain that pollution will increase.

rev. Pollution certainly will increase.

BE FORCEFUL

To develop a forceful, vigorous style, you need to learn some common tricks of the trade and practise them until they become habit. Here are some helpful suggestions.

Choose active over passive verbs

An active verb creates more energy than a passive one does:

> *Active:* Rivers erode valley soils.
> *Passive:* Soils in valleys are eroded by rivers.

The use of passive constructions tends to produce awkward, convoluted phrasing. Writers of bureaucratic documents are among the worst offenders. The passive verbs in the following mouthful make it hard to tell who is doing what:

> It has been decided that the utilization of small rivers in the region for purposes of generating hydro-electric power should be studied by our department and that a report to the Government should be made by our Director as soon as possible.

Passive verbs are appropriate in four instances:

- When the subject is the passive recipient of some action:

  > The politician was heckled by the angry crowd.

- When you want to emphasize the object rather than the person acting:

  > The anti-pollution devices in all three plants will be improved.

- When you want to avoid an awkward shift of subject in a sentence or paragraph. Using the passive sometimes will help you maintain focus:

  > The contractor began to convert the single family homes to apartments but was stopped by the police because a permit had not been issued.

- When you want to avoid placing responsibility or blame:

  > The plans were delayed when the proposer became ill.

When these exceptions don't apply, make an effort to use active verbs for a livelier style.

Use personal subjects

Most of us find it more interesting to learn about people than about things—hence the enduring appeal of the gossip columns. Wherever possible, make the subjects of your sentences personal. This trick goes hand-in-hand with use of active verbs. Almost any sentence becomes more lively with active verbs and a personal subject:

orig.   The outcome of the municipal referendum was the decision to increase the area to be preserved as green space.

rev.   Residents of the municipality voted to increase the area to be preserved as green space.

Here is another example:

orig.   It can be assumed that an agreement was reached, since there were smiles on both city and developer sides when the meeting was finished.

rev.   Apparently the city and the developer reached an agreement, since both bargainers were smiling when they finished the meeting.

Use concrete details

Concrete details are easier to understand—and to remember—than abstract theories. Whenever you are discussing abstract concepts, always provide specific examples and illustrations; if you have a choice between a concrete word and an abstract one, choose the concrete word. Consider this sentence:

The French explored the northern territory and traded with the Native people.

Now see how a few specific details can bring the facts to life:

The French voyageurs paddled their way along the river systems of the north, trading their blankets and copper kettles for the Indians' furs.

Suggesting that you add concreteness doesn't mean getting rid of all abstractions. It's simply a plea to balance the abstractions with accurate details. Additional information, if concrete and accurate, can improve your writing.

Make important ideas stand out

Experienced writers know how to manipulate sentences in order to emphasize certain points. Here are some of their techniques:

PLACE KEY WORDS IN STRATEGIC POSITIONS

The positions of emphasis in a sentence are the beginning and, above all, the end. If you want to bring your point home with force, don't put the key words in the middle of the sentence. Save them for the last:

> **orig.** People are less concerned with efficient and equitable use of resources than they are with getting what they want in this consumption-oriented society.

> **rev.** In this consumption-oriented society, people are less concerned with efficient and equitable use of resources than with getting what they want.

SUBORDINATE MINOR IDEAS

Small children connect incidents with a string of *and*s, as if everything were of equal importance:

> We went to the zoo and we saw a lion and John spilled his drink.

As they grow up, however, children learn to subordinate: that is, to make one part of a sentence less important in order to emphasize another point:

> Because the bus was delayed, we missed our class.

Major ideas stand out more and connections become clearer when minor ideas are subordinated:

> **orig.** The oil well erupted and a repair crew was called in.

> **rev.** When the oil well erupted, a repair crew was called in.

Make your most important idea the subject of the main clause, and try to put it at the end, where it will be most emphatic:

orig. I was relieved when I saw my marks.

rev. When I saw my marks, I was relieved.

VARY SENTENCE STRUCTURE

As with anything else, variety adds spice to writing. One way to add variety, which also will make an important idea stand out, is to use a periodic rather than a simple sentence structure.

Most sentences follow the simple pattern of **s**ubject-**v**erb-**o**bject (plus modifiers):

The <u>dog</u> <u>bit</u> the <u>man</u> on the ankle.
   S   V   O

A *simple sentence* such as this gives the main idea at the beginning of the sentence, making it straightforward and easy to read. A *periodic sentence,* on the other hand, does not give the main clause until the end, following one or more subordinate clauses:

Since he had failed to keep his promises or to inspire the voters, in the next election <u>he</u> <u>was defeated</u>.
       S   V

The longer the periodic sentence is, the greater the suspense and the more emphatic the final part. Since this high-tension structure is more difficult to read than the simple sentence, your readers would be exhausted if you used it too often. Save it for those times when you want to create a special effect or play on emotions.

VARY SENTENCE LENGTH

A short sentence can add punch to an important point, especially when it comes as a surprise. This technique can be used particularly effectively for conclusions. Don't overdo it, though. A string of long sentences may be monotonous, but a string of short ones has a staccato effect that can make your writing sound like a child's reader: "This is my turtle. See him swim." Still, academic papers are more likely to have too many long sentences than too many short ones. Since short sentences are easier to read, try breaking up clusters of long ones. Aim for variety.

USE CONTRAST

Just as a jeweller will highlight a diamond by displaying it against dark velvet, so you can highlight an idea by placing it against a contrasting background:

> **orig.**   Most employees in industry do not have indexed pensions.

> **rev.**   Unlike civil servants, most employees in industry do not have indexed pensions.

Using parallel phrasing will increase the effect of the contrast:

> Although he often spoke to university audiences, he seldom spoke to business groups.

USE A WELL-PLACED ADVERB OR CORRELATIVE CONSTRUCTION

Adding an adverb or two sometimes can help you to dramatize a concept:

> **orig.**   Although I dislike the proposal, I must accept it as the practical answer.

> **rev.**   Although <u>emotionally</u> I dislike the concept, <u>intellectually</u> I must accept it as the practical answer.

Correlatives such as *both . . . and* or *not only . . . but also* can be used to emphasize combinations as well:

> **orig.**   Smith was a good lecturer and a good friend.

> **rev.**   Smith was <u>both</u> a good lecturer <u>and</u> a good friend.

> **rev.**   Smith was <u>not only</u> a good lecturer <u>but also</u> a good friend.

USE REPETITION

Repetition is a highly effective emphatic device. It helps to stir the emotions:

> He fought injustice and corruption. He fought complacent politicians and inept policies. He fought hard, but he always fought fairly.

Of course, you would use such a dramatic technique on rare occasions only.

USE YOUR EARS

Your ears are probably your best critics: make good use of them. Before producing a final copy of any piece of writing, read it out loud, in a clear voice. The difference between cumbersome and fluent passages will be unmistakable.

SOME FINAL ADVICE: WRITE BEFORE YOU REVISE

No one would expect you to sit down and put all this advice into practice as soon as you start to write. You would feel so constrained that it would be hard to get anything down on paper at all. You will be better off if you begin practising these techniques during the editing process, when you are looking critically at what you have already written. Some experienced writers can combine the creative and critical functions, but most of us find it easier to write a rough draft first, before starting the detailed task of revising.

CHAPTER 15

Grammar and Usage

This chapter is not a comprehensive grammar lesson: it's simply a survey of those areas where students most often make mistakes. It will help you to keep a look-out for weaknesses as you are editing your work. Once you get into the habit of checking, it won't be long before you are correcting potential problems as you write. The grammatical terms used here are the most simple and familiar ones; if you need to review some of them, see Chapter 18.

TROUBLES WITH SENTENCE UNITY

Sentence fragments

To be complete, a sentence must have both a subject and a verb in an independent clause; if it does not, it is a fragment. Occasionally a sentence fragment is acceptable, as in

Will the government try to privatize national parks? <u>Not likely</u>.

Here the sentence fragment *not likely* is clearly intended to be understood as a short form of *It is not likely that it will try.* Unintentional sentence fragments, on the other hand, usually seem incomplete rather than shortened:

✗ I enjoy studying geography. Being a person who is interested in the world around me.

The last "sentence" is incomplete: where are the subject and verb? (Remember that a participle such as *being* is not a verb; "-ing" words by themselves are only verbals or part-verbs.) The fragment can be made into a complete sentence by adding a subject and a verb:

✓ <u>I am</u> a person who is interested in the world around me.

Alternatively, you could join the fragment to the preceding sentence:

✓ Being interested in the world around me, I enjoy studying geography.

✓ I enjoy studying geography, since I am interested in the world around me.

Run-on sentences

A run-on sentence is one that continues beyond the point where it should have stopped:

> ✗ Mosquitoes and blackflies are annoying, but they didn't stop tourists from coming to spend their holidays in Canada and such is the case in Ontario's northland.

The *and* should be dropped and a period or semicolon added after *Canada*.

Another kind of run-on sentence is one in which two independent clauses (phrases that could stand by themselves as sentences) are wrongly joined by a comma:

> ✗ Jason Waterman has won international acclaim as a specialist on soil erosion, he is a geography professor at Waitsfield University.

This error is known as a *comma splice*. There are three ways of correcting it:

1. by putting a period after *erosion* and starting a new sentence:

> . . . on soil erosion. He . . .

2. by replacing the comma with a semicolon:

> . . . on soil erosion; he . . .

3. by making one of the independent clauses subordinate to the other:

> ✓ Jason Waterman, who has won international acclaim as a specialist on soil erosion, is a geography professor at Waitsfield University.

The one exception to the rule that independent clauses cannot be joined by a comma arises when the clauses are very short and arranged in a tight sequence:

> I opened the door, I saw the skunk, I shut the door.

Such instances are uncommon, obviously.

Contrary to what many people think, words such as *however*, *therefore*, and *thus* cannot be used to join independent clauses:

> ✗ Two of my friends started out in Commerce, however they quickly decided they didn't like accounting.

The mistake can be corrected by beginning a new sentence after *commerce* or (preferably) by putting a semicolon in the same place:

> ✓ Two of my friends started out in Commerce; however, they quickly decided they didn't like accounting.

The only words that can be used to join independent clauses are the coordinating conjunctions—*and*, *or*, *nor*, *but*, *for*, *yet*, and *so*—and subordinating conjunctions such as *if*, *because*, *since*, *while*, *when*, *where*, *after*, *before*, and *until*.

Faulty predication

When the subject of a sentence is not connected grammatically to what follows (the predicate), the result is faulty predication:

> ✗ The <u>reason</u> for the failure of the IMF program was <u>because</u> it did not address the needs of the poor.

The problem here is that *because* means essentially the same thing as *the reason for*. The subject needs a noun clause to complete it:

> ✓ The <u>reason</u> for the failure of the IMF program was <u>that</u> it did not address the needs of the poor.

Another solution would be to rephrase the sentence:

> ✓ The IMF program failed because it did not address the needs of the poor.

Faulty predication also occurs with *is when* and *is where* constructions:

> ✗ The critical moment <u>is when</u> the researcher discovers the original source of the information.

You can correct this error in one of two ways:

1. Follow the *is* with a noun phrase to complete the sentence:

> ✓ The critical moment is <u>the discovery of the original source</u> by the researcher.

(or)

✓ The critical moment is <u>the researcher's discovery</u> of the original source.

2. Change the verb:

✓ The critical moment <u>occurs</u> when the researcher discovers the original source.

TROUBLES WITH SUBJECT–VERB AGREEMENT

Identifying the subject

A verb should always agree in number with its subject. Sometimes, however, when the subject does not come at the beginning of the sentence, or when it is separated from the verb by other information, you may be tempted to use a verb form that does not agree:

✗ The <u>increase</u> in the rate for freight and passengers <u>were condemned</u> by the farmers.

The subject here is *increase*, not *freight and passengers*; therefore the verb should be the singular *was condemned*:

✓ The <u>increase</u> in the rate for freight and passengers <u>was condemned</u> by the farmers.

Either, neither, each

The indefinite pronouns *either*, *neither*, and *each* always take singular verbs:

✗ <u>Neither</u> of those glaciers <u>have</u> a major crevasse.

✓ <u>Each</u> of these glaciers <u>has</u> many crevasses.

Compound subjects

When *or*, *either . . . or*, or *neither . . . nor* is used to create a compound subject, the verb usually should agree with the last item in the subject:

✓ Neither the city planner nor <u>the developers are going</u> to the Municipal Board hearing.

If a singular item follows a plural item, however, a singular verb may sound awkward, and it's better to rephrase the sentence:

orig.   Either my history <u>books</u> or my biology <u>text</u> <u>is going</u> to gather dust this weekend.

rev.   This weekend, I am going to read either my history books or my biology text.

Unlike the word *and*, which creates a compound subject and therefore takes a plural verb, *as well as* or *in addition to* does not create a compound subject; therefore the verb remains singular:

✓ Fertile soil <u>and</u> a good climate <u>are</u> important requirements for agriculture.

✓ Fertile soil <u>as well as</u> a good climate <u>is</u> an important requirement for agriculture.

Collective nouns

A collective noun is a singular noun, such as *family*, *class*, *army*, or *team*; this includes a number of members. If the noun refers to the members as a unit, it takes a singular verb:

✓ The cultural geography <u>class</u> <u>goes</u> to the museum each year.

If the noun refers to the members as individuals, however, the verb becomes plural:

✓ The <u>team are receiving</u> their sweaters before the exhibition game.

✓ The <u>majority</u> of immigrants to North America <u>settle</u> in cities.

Titles

A title is singular even if it contains plural words; therefore it takes a singular verb:

✓ *Tales of the South Pacific* was a best-seller.

✓ McCarthy and McCarthy is handling the court case.

TENSE TROUBLES

Native speakers of English usually know the correct sequence of verb tense by ear, but a few tenses can still be confusing.

The past perfect

If the main verb is in the past tense and you want to refer to something before that time, use the past perfect (*had* plus the past participle). The time sequence will not be clear if you use the simple past in both clauses:

✗ He hoped that she finished reading the map.

✓ He hoped that she had finished reading the map.

Similarly, when you are reporting what someone said in the past—that is, when you are using past indirect discourse—you should use the past perfect tense in the clause describing what was said:

✗ He said that the tractor caused the soil compaction.

✓ He said that the tractor had caused the soil compaction.

Using "if"

When you are describing a possibility in the future, use the present tense in the condition (*if*) clause and the future tense in the consequence clause:

✓ If he tests us on Hägerstrand's diffusion model, I shall fail.

When the possibility is unlikely, it is conventional—especially in formal writing—to use the subjunctive in the *if* clause, and *would* plus the base verb in the consequence clause:

✓ If she were to cancel the exam, I would cheer.

When you are describing a hypothetical instance in the past, use the past subjunctive (it has the same form as the past perfect) in the *if* clause, and *would have* plus the past participle for the consequence. A common error is to use *would have* in both clauses:

✗ If he would have been clearer, I would have been able to understand the question.

✓ If he had been clearer, I would have been able to understand the question.

Writing about literature

When you are describing a literary work in its historical context, use the past tense:

✓ Margaret Atwood <u>wrote</u> *Surfacing* at a time when George Grant's *Technology and Empire* <u>was persuading</u> people to reassess technocratic values.

To discuss what goes on *within* a work of literature, however, you should use the present tense:

✓ The narrator <u>retreats</u> to the woods and <u>tries</u> to escape the rationalism of her father's world.

When you are discussing an episode or incident in a literary work and want to refer to a prior incident or a future one, use past or future tenses accordingly:

✓ The narrator returns to Otago, where she <u>spent</u> her summers as a child; by the time she leaves, she <u>will have rediscovered</u> herself.

Be sure to return to the present tense when you have finished referring to events in the past or future.

PRONOUN TROUBLES

Pronoun reference

The link between a pronoun and the noun it refers to must be clear. If the noun doesn't appear in the same sentence as the pronoun, it should appear in the preceding sentence:

✗ The textbook supply in the bookstore had run out, and so we borrowed <u>them</u> from the library.

Since *textbook* is used as an adjective rather than a noun, it cannot serve as referent or antecedent for the pronoun *them*. You must either replace *them* or change the phrase *textbook supply*.

✓ The <u>textbook supply</u> in the bookstore had run out, and so we borrowed the <u>texts</u> from the library.

✓ The <u>textbooks</u> in the bookstore had run out, and so we borrowed <u>them</u> from the library.

When a sentence contains more than one noun, make sure there is no ambiguity about which noun the pronoun refers to:

✗ The public wants increased social <u>services</u> as well as lower <u>taxes</u>, but the government does not advocate <u>them</u>.

What does the pronoun *them* refer to? The taxes, the social services, or both?

✓ The public wants <u>increased</u> social <u>services</u> as well as lower taxes, but the government does not advocate spending <u>increases</u>.

Using "it" and "this"

Using *it* and *this* without a clear referent can lead to confusion:

✗ Although the directors wanted to meet in January, <u>it</u> (<u>this</u>) didn't take place until May.

✓ Although the directors wanted to meet in January, <u>the conference</u> didn't take place until May.

Make sure that *it* or *this* clearly refers to a specific noun or pronoun.

Pronoun agreement and gender

A pronoun should agree in number and person with the noun that it refers to. Traditionally, the following sentence would have been considered incorrect:

When a federal <u>employee</u> retires, <u>their</u> pension is indexed.

In the past, this sentence would have been revised as follows:

When a federal <u>employee</u> retires, <u>his</u> pension is indexed.

The problem with this revision is that, in using the pronoun *his*, it appears to neglect the many female federal employees who also retire. Although some language experts still maintain that *he* has a dual meaning, one for an individual male and one for any human, today this usage is widely regarded as sexist. One solution is to use *his/her*, but this phrase is intrusive and awkward. For that reason, use of the plural *their* with reference to a singular noun, as in the first example above, is becoming increasingly common, and this trend appears to be gaining acceptance. Some people still object to it, however. To be on the safe side, you may prefer to rephrase the sentence—for example, by changing the singular noun to a plural:

When federal <u>employees</u> retire, <u>their pensions are</u> indexed.

Whatever form you choose, check for agreement between subjects and verbs. Use neutral nouns whenever possible (see pp. 207–8). And, where appropriate, at least try to make clear in your examples and illustrations that you are referring to females as well as males—unless there is a clear need to differentiate.

Using "one"

People often use the word *one* to avoid over-using *I* or *you* in their writing. Although in Britain this is common, in Canada and the United States frequent use of *one* may seem too formal and even a bit pompous:

If <u>one</u> were to apply for the grant, <u>one</u> would find oneself engulfed in so many bureaucratic forms that <u>one's</u> patience would be stretched thin.

In the past, a common way around this problem was to use the third person *his* or *her* as the adjectival form of *one*. Yet today this usage is increasingly regarded as unacceptable. As we saw in the preceding section, in some cases you may be able to substitute the plural *their*; just remember that some people still object to this usage as well.

In any case, try to use *one* sparingly, and don't be afraid of the occasional *I*. The one serious error is to mix the third person *one* with the second person *you*:

✗   When <u>one</u> visits the Rockies, <u>you</u> are impressed by the grandeur of the scenery.

In formal academic writing generally, *you* is not an appropriate substitute for *one*.

Using "me" and other objective pronouns

Remembering that it is wrong to say "Ahmed and *me* were invited to speak to the planning committee" rather than "Ahmed and *I* were invited," many people use the subjective form of the pronoun even when the objective form is correct:

✗   The planning committee invited Ahmed and <u>I</u> to speak.

✓   The planning committee invited Ahmed and <u>me</u> to speak.

The verb *invited* requires an object, and *me* is the objective case. Here is a simple hint: read the sentence with only the problem pronoun. You will know by ear which form is correct:

✓ The planning committee invited <u>me</u> to speak.

Prepositions should also be followed by the objective case:

✗ <u>Between</u> you and <u>I</u>, Brown is a bore.

✓ <u>Between</u> you and <u>me</u>, Brown is a bore.

✗ Eating well is a problem <u>for we</u> students.

✓ Eating well is a problem <u>for us</u> students.

There are times, however, when the correct case can sound stiff or awkward:

To whom was the award given?

Rather than keep to a correct but awkward form, try to reword the sentence:

Who received the award?

Exceptions for pronouns following prepositions
The rule that a pronoun following a preposition takes the objective case has exceptions. When the preposition is followed by a clause, the pronoun should take the case required by its position in the clause:

✗ The students showed some concern over <u>whom would be selected</u> as spokesperson.

Although the pronoun follows the preposition *over*, it is also the subject of the verb *would be selected* and therefore requires the subjective case:

✓ The students showed some concern over <u>who would be selected</u> as spokesperson.

Similarly, when a gerund (a word that acts partly as a noun and partly as a verb) is the subject of a clause, the pronoun that modifies it takes the possessive case:

✗ The professor was impressed by <u>him presenting</u> both sides of the issue.

✓ The professor was impressed by <u>his presenting</u> both sides of the issue.

✗ He was tired of <u>me reminding</u> him.

✓ He was tired of <u>my reminding</u> him.

TROUBLES WITH MODIFYING

Adjectives modify nouns; adverbs modify verbs, adjectives, and other adverbs. Do not use an adjective to modify a verb:

✗ He played <u>good</u>. (Adjective with verb)

✓ He played <u>well</u>. (Adverb modifying verb)

✓ He played <u>really well</u>. (Adverb modifying adverb)

✓ He had a <u>good style</u>. (Adjective modifying noun)

✓ He had a <u>really good style</u>. (Adverb modifying adjective)

Squinting modifiers

Remember that clarity largely depends on word order: to avoid confusion, the relations between the different parts of a sentence must be clear. Modifiers should be as close as possible to the words they modify. A *squinting modifier* is one that, because of its position, seems to look in two directions at once:

✗ She expected <u>in the spring</u> the river to flood the valley.

Was *spring* the time of expectation or the time of the flood? To make the logical relation clear, try changing the order of the sentence, or rephrasing:

✓ <u>In the spring</u> she <u>expected</u> the river to flood the valley.

✓ She expected the river <u>to flood</u> the valley <u>in the spring</u>.

✓ She expected a <u>spring flood</u> in the valley.

Other squinting modifiers can be corrected in the same way:

✗ Our professor gave a lecture on rock slides in Mellon Hall, <u>which was well illustrated</u>.

✓ Our professor gave a <u>well-illustrated lecture</u> in Mellon Hall on <u>rock slides</u>.

Often the modifier works best when it is placed immediately in front of the element it modifies. Notice the difference that this placement makes:

Only she guessed the source of the error.

She only guessed the source of the error.

She guessed only the source of the error.

Dangling modifiers

Modifiers that have no grammatical connection with anything else in the sentence are said to be *dangling*:

✗ Walking around the campus in July, the river and trees made a picturesque scene.

Who is doing the walking? Here's another example:

✗ Reflecting on the results of the referendum, it was decided not to press for independence for a while.

Who is doing the reflecting? Clarify the meaning by connecting the dangling modifier to a new subject:

✓ Walking around the campus in July, she thought the river and trees made a picturesque scene.

✓ Reflecting on the results of the referendum, the government decided not to press for independence for a while.

TROUBLES WITH PAIRS (AND MORE)

Comparisons

Make sure that your comparisons are complete. The second element in a comparison should be equivalent to the first, whether the equivalence is stated or merely implied:

✗ Today's students have a greater understanding of calculus than their parents.

This sentence suggests that the two things being compared are *calculus* and *parents*. Adding a second verb (*have*) equivalent to the first one shows that the two things being compared are *parents' understanding* and *students' understanding*:

> ✓ Today's students <u>have</u> a greater understanding of calculus than their parents <u>have</u>.

A similar problem arises in the following comparison:

> ✗ That politician is <u>a tiresome man</u> and so are his press conferences.

Press conferences may be tiresome, but they are not a tiresome man. To make sense, the two parts of the comparison must be parallel:

> ✓ That politician is <u>tiresome</u>, and so are his press conferences.

Correlatives (coordinate constructions)

Constructions such as *both . . . and, not only . . . but*, and *neither . . . nor* are especially tricky. The coordinating term must not come too early, or else one of the parts that follows will not connect with the common element. For the implied comparison to work, the two parts that come after the coordinating term must be grammatically equivalent:

> ✗ He <u>not only</u> bakes cakes <u>but also</u> bread.

> ✓ He bakes <u>not only</u> cakes <u>but also</u> bread.

Parallel phrasing

A series of items in a sentence should be phrased in parallel wording. Make sure that all the parts of a parallel construction are in fact equal:

> ✗ Mackenzie King loved <u>his</u> job, <u>his</u> dogs, and mother.

> ✓ Mackenzie King loved <u>his</u> job, <u>his</u> dogs, and <u>his</u> mother.

Once you have decided to include the pronoun *his* in the first two elements, the third must have it too.

For clarity as well as stylistic grace, keep similar ideas in similar form:

> ✗ He <u>failed</u> Economics and <u>barely passed</u> Statistics, but Periglacial Geomorphology <u>was</u> a subject he did well in.

> ✓ He <u>failed</u> Economics and <u>barely passed</u> Statistics, but <u>did well</u> in Periglacial Geomorphology.

CHAPTER 16

Punctuation

Punctuation causes students so many problems that it deserves a chapter of its own. If your punctuation is faulty, your readers will be confused and may have to backtrack; worse still, they may be tempted to skip over the rough spots. Punctuation marks are the traffic signals of writing; use them with precision to keep readers moving smoothly through your work. Items in this chapter are arranged alphabetically: *apostrophe, brackets, colon, comma, dash, ellipsis, exclamation mark, hyphen, italics or underlining, parentheses, period, quotation marks, quotation marks with punctuation,* and *semicolon.*

APOSTROPHE [']

The apostrophe forms the possessive case for nouns and some pronouns. (Remember that possession may also be shown without an apostrophe if the possessor is preceded by *of,* as in *the work of McLuhan,* or *the end of the day.*)

1. **Add an apostrophe followed by s to**

   - all singular and plural nouns not ending in *s: the cat's dish, women's studies.*
   - singular *proper* nouns ending in *s: Keats's poetry, Sis's birthday* (but note that the final *s* can be omitted if the word has a number of them already and would sound awkward, as in *Jesus'* or certain classical names)
   - indefinite pronouns: *someone's, anybody's,* etc.

2. **Add an apostrophe to plural nouns ending in s**: *our families' houses, the players' uniforms.*

3. **Use an apostrophe to show contractions of words**: *isn't, can't, winter of '97.* Caution: don't confuse *it's* (the contraction of *it is*) with the possessive of *it* (*its*), which has no apostrophe.

BRACKETS [ ]

Brackets are square enclosures, not to be confused with parentheses (which are round).

1. **Use brackets to set off a remark of your own within a quotation.** They show that the words enclosed are not those of the person quoted:

    > Cowan maintains, "Obstacles to western unification [in the next decade] are as many as they are serious."

    Brackets are sometimes used to enclose *sic* (Latin for *thus*), which is used after an error, such as a misspelling, to show that the mistake was in the original. *Sic* may be italicized or underlined:

    > The politician, in his letter to constituents, wrote about "these parlouse [*sic*] times of economic difficulty."

COLON [:]

A colon indicates that something is to follow.

1. **Use a colon before a formal statement or series:**

    ✓ The winners are the following: Alain, Vladimir, and Sonia.

    Don't use a colon if the words preceding it do not form a complete sentence:

    ✗ The winners are: Alain, Vladimir, and Sonia.

    ✓ The winners are Alain, Vladimir, and Sonia.

    Occasionally, however, a colon is used if the list is arranged vertically:

    ✓ The winners are: Alain
    Vladimir
    Sonia

2. **Use a colon for formality before a direct quotation:**

    > The leaders of the anti-nuclear group repeated their message: "The world needs bread before bombs."

3. **Use a colon between numbers or groups of numbers expressing time and ratios:**

> 5:46 p.m.

> 1:100 (as in "1 cm equals 100 m", or "1 inch equals 100 feet")

COMMA [,]

Commas are the trickiest of all punctuation marks; even the experts differ on when to use them. Most agree, however, that too many commas are as bad as too few, since they make writing choppy and awkward to read. Certainly recent writers use fewer commas than earlier stylists did. Whenever you are in doubt, let clarity be your guide. The most widely accepted conventions are these:

1. **Use a comma to separate two independent clauses joined by a coordinating conjunction (and, but, for, or, nor, yet, so).** By signalling that there are two clauses, the comma will prevent readers from confusing the beginning of the second clause with the end of the first:

✗ He went out for dinner with his sister and his roommate joined them later.

✓ He went out for dinner with his sister, and his roommate joined them later.

When the second clause has the same subject as the first, you have the option of omitting both the second subject and the comma:

✓ He can construct a stem and leaf plot, but he cannot calculate the mean.

✓ He can construct a stem and leaf plot but cannot calculate the mean.

If you mistakenly punctuate two sentences as if they were one, the result will be a *run-on sentence*; if you use a comma but forget the coordinating conjunction, the result will be a *comma splice*:

✗ She went to the library, it was closed.

✓ She went to the library, but it was closed.

Remember that words such as *however*, *therefore*, and *thus* are conjunctive adverbs, not conjunctions: if you use one of them the way you would use a conjunction, the result will again be a *comma splice*:

✗ She was accepted into graduate school, however, she took a year off to earn her tuition.

✓ She was accepted into graduate school; however, she took a year off to earn her tuition.

Conjunctive adverbs often are confused with conjunctions. You can distinguish between the two if you remember that a conjunctive adverb's position in a sentence can be changed:

✓ She was accepted into graduate school; she took a year off, however, to earn her tuition.

The position of a conjunction, on the other hand, is invariable; it must be placed between the two clauses:

✓ She was accepted into graduate school, but she took a year off to earn her tuition.

When, in rare cases, the independent clauses are short and closely related, they may be joined by a comma alone:

✓ I came, I saw, I conquered.

A *fused sentence* is a run-on sentence in which independent clauses are slapped together with no punctuation at all:

✗ He watched the wave tank all afternoon the only exercise he got was going to the cafeteria for an occasional break.

A fused sentence sounds like breathless babbling—and it is a serious error.

2. **Use a comma between items in a series.** Place a coordinating conjunction before the last item:

✓ The landscape had nucleated villages, houses with mansard roofs, and "French" church architecture.

✓ Continuing agricultural expansion in frontier areas is a reflection of economic factors, technological change, available labour, and government policies.

The comma before the conjunction is optional:

✓ She kept a cat, a dog and a budgie.

Sometimes, however, the final comma can help to prevent confusion:

✓ When we set out to do our field work, we were warned about muddy fields, angry farmers, trespassing on private property, and bears.

In this case, the comma prevents the reader from thinking that "we" might trespass on *bears* as well as on private property.

3. **Use a comma to separate adjectives preceding a noun when they modify the same element:**

It was a rainy, windy night.

When the adjectives do *not* modify the same element, you should *not* use a comma:

It was a pleasant winter outing.

Here *winter* modifies *outing*, but *pleasant* modifies the total phrase *winter outing*. A good way of determining whether or not you need a comma is to see whether you can reverse the order of the adjectives. If you can reverse it (*rainy, windy night* or *windy, rainy night*), use a comma; if you cannot (*winter pleasant outing*), omit the comma.

4. **Use commas to set off an interruption (or "parenthetical element"):**

✓ The film, I hear, isn't nearly as good as the book.

✓ My tutor, however, couldn't answer the question.

Remember to put commas on both sides of the interruption:

✗ The model, they say was adapted from a model in biology.

✓ The model, they say, was adapted from a model in biology.

5. **Use commas to set off words or phrases that provide additional but non-essential information:**

✓ Our president, Sue Stephens, does her job well.

✓ The black retriever, his closest companion, went with him everywhere.

Sue Stephens and *his closest companion* are *appositives:* they give additional information about the nouns they refer to (*president* and *retriever*), but the sentences would be understandable without them. Here is another example:

✓ The Minister of Environment, who represents Edmonton South, was in Paris last week.

The phrase *who represents Edmonton South* is called a *non-restrictive* modifier, because it does not limit the meaning of the phrase it modifies (*Minister of Environment*). Without that modifying clause the sentence would still specify who was in Paris last week. Since the information the clause provides is not necessary to the meaning of the sentence, you must use commas on both sides to set it off.

By contrast, a *restrictive* modifier is one that provides essential information; it must not be set apart from the element it modifies, and commas should not be used:

✓ The Six Nations people who live near Brantford do not yet hold title to land along the entire Grand River.

Without the clause *who live near Brantford*, the reader might not know where the people live.

To avoid confusion, be sure to distinguish carefully between essential and additional information. The difference can be important:

> Students, who are not willing to work, should not receive grants.

> Students who are not willing to work should not receive grants.

6. **Use a comma after an introductory phrase when omitting it would cause confusion:**

✗ On the river bank above the happy fluvial geomorphologist stood.

✓ On the river bank above, the happy fluvial geomorphologist stood.

✗ When he turned away some students quietly left the room.

✓ When he turned away, some students quietly left the room.

7. **Use a comma to separate elements in dates and addresses:**

> February 2, 2007 (commas often are omitted if the day comes first: "2 February 2007")

117 Hudson Drive, Edmonton, Alberta.

They lived in Dartmouth, Nova Scotia.

Note that in Britain a comma is placed after the number in an address: "117, Hudson Drive."

8. **Use a comma before a quotation in a sentence:**

   He said, "Life is too short to worry."

   "The children's safety," he warned, "is in your hands."

For more formality, you may use a colon (see p. 244).

9. **Use a comma with a name followed by a title:**

   D. Gunn, Ph.D.

   Alice Smith, President

10. **Do not use a comma between a subject and its verb:**

    ✗ The editor of the book, is Miriam Stoyanovich.

    ✓ The editor of the book is Miriam Stoyanovich.

11. **Do not use a comma between a verb and its object:**

    ✗ Felber said, that the soil sample was large enough for the experiment.

    ✓ Felber said that the soil sample was large enough for the experiment.

12. **Do not use a comma between a coordinating conjunction and following clause:**

    ✗ Go to the lab now or, once you complete the test.

    ✓ Go to the lab now or once you complete the test.

DASH [— OR --]

A dash creates an abrupt pause, emphasizing the words that follow. (Most computer programs today can produce dashes. If yours can't, type two hyphens together, with no space on either side. Both forms are shown here.)

Never use dashes as casual substitutes for other punctuation: overuse can detract from the calm, well-reasoned effect you want.

1. **Use a dash to stress a word or phrase:**

    The British--as a matter of honour--vowed to retake the islands.

    Foster was well received in the office—at first.

2. **Use a dash in interrupted or unfinished dialogue:**

    "It's a matter—to put it delicately—of personal hygiene."

    "But I thought—" Donald tried to explain, but Mario cut him off: "You were wrong."

ELLIPSIS [. . .]

1. **Use an ellipsis (three spaced dots) to show the location of any omission from a quotation:**

    -   If you start a quotation in the middle of a sentence, indicate the omission of the earlier material by inserting an ellipsis (*three* dots):

        . . . a wood stove does not burn efficiently at very low temperatures.

    -   If you quote two parts of one sentence, link them with *three* dots. Leave one space after the last word before the ellipsis and another before the word following it:

        A wood stove . . . creates good heat when used carefully.

    -   If you omit the end of the sentence, indicate this by using *four* dots (the first one represents the period at the end of the sentence):

        A wood stove made from cast iron creates good heat. . . .

    -   Four dots followed by a new sentence indicate only that some material has been omitted, whether the end of the first sentence or one or more complete sentences following it. Don't forget to leave a space between the last dot and the new sentence:

        A wood stove made from cast iron creates good heat when used carefully. . . . Oak has high heat value.

2. **Use an ellipsis to show that a series of numbers continues indefinitely:**

    1, 3, 5, 7, 9 . . .

EXCLAMATION MARK [!]

An exclamation mark helps to show emotion or feeling. It usually is found in dialogue:

    "Woe is me!" she mourned.

In academic writing, you should use an exclamation mark only in those rare cases when you want to give a point an emotional emphasis:

    He concluded that inflation would decrease in 1991. Some forecast!

HYPHEN [-]

1. **Use a hyphen if you must divide a word at the end of a line.** When a word is too long to fit at the end of a line, try to keep it in one piece by starting a new line. If you must divide, however, remember these rules:

   • Divide between syllables.
   • Never divide a one-syllable word.
   • Never leave one letter by itself.
   • Divide double consonants except when they come before a suffix, in which case divide before the suffix:

        ar-rangement

        embar-rassment

        fall-ing

        pass-able

When the second consonant has been added to form the suffix, keep it with the suffix:

        begin-ning

        refer-ral

2. **Use a hyphen to separate the parts of certain compound words:**

- compound nouns:

  sister-in-law, Lieutenant-Governor, Vice-President.

- compound verbs:

  test-drive, over-simplify.

- compound adjectives used as modifiers preceding nouns:

  well-considered plan, twentieth-century attitudes.

When you are not using such expressions adjectivally, do not hyphenate them:

The plan was well considered.

These are attitudes of the twentieth century.

After long-time use, some compound nouns drop the hyphen. When in doubt, check a dictionary.

3. **Use a hyphen with certain prefixes** (*all-*, *self-*, *ex-*, and those prefixes preceding a proper name):

   all-party, self-imposed, ex-jockey, anti-nuclear, pro-Trinidad

4. **Use a hyphen to emphasize contrasting prefixes:**

   The coach agreed to give both pre- and post-game interviews.

5. **Use a hyphen to separate written-out compound numbers from one to one hundred and compound fractions used as modifiers:**

   eighty-one years ago

   seven-tenths full

6. **Use a hyphen to separate parts of inclusive numbers or dates:**

   the years 1890-1914

   pp. 3-40

ITALICS [*ITALICS*] OR UNDERLINING

Italics are slanted (cursive) letters. While italics are much more commonly used, some instructors may require that book and journal titles be underlined. If you are writing something in long-hand, remember to underline where necessary, as free hand italics are usually indistinguishable in handwriting.

1. **Use italics (or underlining) for the titles of books, long poems that are complete books, plays, films, and lengthy musical pieces:**

   Yi-Fu Tuan's *Space and Place* is one of my favourite books.

For articles, essays, and short poems or musical pieces, use quotation marks. If the title contains another title, be sure to set it off in the correct style:

- When both titles are books, use quotation marks for the internal one:

  Her latest book is *An Interpretation of "Space and Place."*

- When the internal title is a book but the main title is not, use italics (or underlining):

  For more detail see Joseph Thrill's essay "Lawrence Kaplan's *Street Kids and Political Reaction* as a Primer for Understanding Poverty and Despair," in Agard Sinclair's edited book *Humanistic Geography of the City.*

- When neither title is a book, use the alternative form of quotation marks:

  My essay is entitled "Imagery in 'Whale Songs' by Evans."

2. **Use italics (or underlining) to emphasize an idea:**

   It is important that all equipment be washed *immediately.*

Be sparing with this use of italics. For less intrusive ways of creating emphasis see pp. 226–9.

3. **Use italics (or underlining) to emphasize a word (or phrase):**

   The term *areal differentiation* is an example of geographical jargon.

Quotation marks may be used for the same purpose.

4. **Use italics (or underlining) for scientific names:**

> Cape May Warbler (*Dendroica tigrina*)
>
> Black-throated Green Warbler (<u>Dendroica virens</u>)
>
> *Pinus radiata* (a species of pine)

PARENTHESES [( )]

1. **Use parentheses to enclose an explanation, example, or qualification.** Parentheses show that the enclosed material is of incidental importance to the main idea. They make a less pronounced interruption than a dash, but a more pronounced one than a comma:

> My wife (who is the mother of our two children) is a great plumber.
>
> His latest plan (according to neighbours) is to dam the creek.

Remember that although punctuation should not precede parentheses, it may follow them if required by the sense of the sentence:

> I like coffee in the morning (if it's not instant), but she prefers tea.

If the parenthetical statement comes between two complete sentences, it should be punctuated as a sentence, with the period inside the parentheses:

> I finished my last essay on April 30. (It was on Aristotle's ethics.)
> Fortunately, I had three weeks free to study for the exam.

2. **Use parentheses to enclose references.** See Chapter 11 for details.

PERIOD [.]

1. **Use a period at the end of a sentence.** A period indicates a full stop, not just a pause.

2. **Use a period with abbreviations.** Whereas British style omits the period in certain cases (UK, US, BSc, MA, PhD, Dr, Revd, etc.), North American style usually requires it for abbreviated titles (B.Sc., M.A., Ph.D., Dr., Rev., etc.) as well as place-names (U.K., U.S., B.C., N.W.T., N.Y., VT., etc.) and

expressions of time (6:30 p.m.); note, however, that no periods are used for place-names in mailing addresses (ON, NS, NY, VT). While the acronyms and initialisms for some organizations include periods, the most common ones generally do not (BBC, CIA, CIDA, FAO, IDRC, MIT, RCMP, UNESCO, etc.). Some (e.g., UN [or U.N.]) appear with and without periods. Whichever format you use, be consistent.

3. **Use a period at the end of an indirect question.** Do not use a question mark:

✗ He asked if I wanted a substitute?

✓ He asked if I wanted a substitute.

4. **Use a period for questions that are really polite orders:**

Will you please send him the report by Friday.

QUOTATION MARKS [" " OR ' ']

Quotation marks are usually double in American style and single in British. In Canada either is accepted—just be consistent.

1. **Use quotation marks to signify direct discourse (the actual words of a speaker):**

I asked, "What is the matter?"

He said, "I have a pain in my big toe."

2. **Use quotation marks to show that words themselves are the issue:**

The term "love" in tennis comes from the French word for "egg."

Alternatively, you may italicize or underline the terms in question.

Sometimes quotation marks are used to mark a slang word or an inappropriate usage, to show that the writer is aware of the difficulty:

The "experts" were wrong.

Use this device only when necessary. In general, it is better to let the context show your attitude, or to choose another term.

3. **Use quotation marks to enclose the titles of poems, short stories, songs, and articles in books or journals.** By contrast, titles of books, films, paintings, or longer musical works are italicized or underlined:

> The essay I liked best in James Smith's *Agricultural Land: A Compendium* is "The Rural-Urban Fringe" by Z. Oseia.

4. **Use quotation marks to enclose quotations within quotations** (single or double, depending on your primary style):

> He said, "Several of the 'expert witnesses' knew nothing about the topic."

5. **Do not use quotation marks for indented, single-spaced block quotations.** When the material being quoted is *four lines or longer*, it should be indented (usually with a double indent) and single-spaced. No quotation marks should be used. If the block quotation is from the beginning of a paragraph, the normal indentation of the first word should be included:

> If the isolated village of Bytown ever becomes the capital of the country, it is clear that thereafter not only will Bytown grow but all the occupied lands will speedily be settled by a sturdy and industrious people. Hamlets will spring, as if by magic, into villages, villages into towns, and towns into cities.

If the material is not from the beginning of a paragraph, no extra indentation is required.

QUOTATION MARKS WITH PUNCTUATION

Both the British and the American practices are accepted in Canada. British style usually places the punctuation outside the quotation marks, unless it is actually part of the quotation. The American practice, followed in this book, is increasingly common in Canada:

- A comma or period always goes inside the quotation marks:

> He said, "Give me another chance," but I replied, "You've had enough chances."

- A semicolon or colon always goes outside the quotation marks:

> Sandy wants to watch "The Journal"; I'd rather watch the rugby game.

- A question mark, dash, or exclamation mark goes inside quotation marks if it is part of the quotation, but outside if it is not:

  He asked, "What's for dinner?"

  Did he really call the boss a "lily-livered hypocrite"?

  His speech was hardly an appeal for "blood, sweat and tears"!

  I was just whispering to Mark, "That instructor is a —" when suddenly she glanced at me.

- When a reference is given parenthetically (in round brackets) at the end of a quotation, the quotation marks precede the parentheses and the sentence punctuation follows them, and there is no period before the source information:

  Riverton suggests that we should "abandon spring burning as a means for fertilizing soils" (*Daily Rag*, 12 April 2007).

SEMICOLON [;]

A semicolon indicates a degree of separation intermediate in value between a comma and a period.

1. **Use a semicolon to join independent clauses (complete sentences) that are closely related:**

   For five days he worked non-stop; by Saturday he was exhausted.

   His lecture was confusing; no one could understand the terminology.

A semicolon is especially useful when the second independent clause begins with a conjunctive adverb such as *however, moreover, consequently, nevertheless, in addition,* or *therefore* (usually followed by a comma):

   He bought a bag of doughnuts; however, none of the group was hungry.

Some grammarians may disagree, but it usually is acceptable to follow a semicolon with a coordinating conjunction if the second clause is complicated by other commas:

   Blair, my cousin, is a keen jogger in all weather; but sometimes, especially in winter, I think it does him more harm than good.

2. **Use a semicolon to mark the divisions in a complicated series when individual items themselves need commas.** Using a comma to mark the subdivisions and a semicolon to mark the main divisions will help to prevent mix-ups:

> ✗　He invited Shiraz Habib, the vice-principal, Jane Hunter, and John Jenkins.

Is the vice-principal a separate person? If so, changing all the commas to semicolons will make that clear. On the other hand, if either Habib or Hunter is the vice-principal, one comma should remain; its location will depend on which of the two the title refers to:

> ✓　He invited Shiraz Habib, the vice-principal; Jane Hunter; and John Jenkins.

> ✓　He invited Shiraz Habib; the vice-principal, Jane Hunter; and John Jenkins.

In a case such as this, the elements separated by the semicolon need not be independent clauses.

UNDERLINING

See "Italics," pp. 253–4.

CHAPTER 17

Misused Words and Phrases

Here are some words and phrases that often are misused. A periodic read-through will refresh your memory and help you avoid needless mistakes.

accept, except. Accept is a verb meaning to *receive affirmatively;* **except,** when used as a verb, means to exclude:

> I accept your offer.

> The teacher excepted him from the general punishment.

accompanied by, accompanied with. Use **accompanied by** for people; **accompanied with** for objects:

> He was accompanied by his wife.

> The brochure arrived, accompanied with a discount coupon.

advice, advise. Advice is a noun, **advise** a verb:

> He was advised to ignore the others' advice.

affect, effect. Affect is a verb meaning to *influence;* **effect** can be either a noun meaning *result* or a verb meaning to *bring about.*

> The eye drops affect his vision.

> The effect of higher government spending is higher inflation.

all ready, already. To be **all ready** is simply to be ready for something; **already** means *beforehand* or *earlier:*

> The students were all ready for the lecture to begin.

> The professor had already left her office by the time Blair arrived.

all right. Write as two separate words: *all right*. This can mean *safe and sound, in good condition, okay*; *correct*; *satisfactory*; or *I agree*.

> Are you all right?

> The student's answers were all right.

(But does the last mean the answers were all correct, or simply satisfactory?)

all together, altogether. All together means *in a group;* **altogether** is an adverb meaning *entirely*:

> He was altogether certain that the children were all together.

allusion, illusion. An **allusion** is an indirect reference to something; an **illusion** is a false perception:

> The rock image is an allusion to the myth of Sisyphus.

> He thought he saw a sea monster, but it was an illusion.

a lot. Write as two separate words: *a lot*.

alternate, alternative. Alternate means *every other* or *every second* thing in a series; **alternative** refers to a *choice* between options:

> The two sections of the class attended discussion groups on alternate weeks.

> Is there an alternative route to the park?

among, between. Use **among** for three or more persons or objects, **between** for two:

> Between you and me, there's trouble among the team members.

amount, number. Amount indicates quantity when units are not discrete and not absolute; **number** indicates quantity when units are discrete and absolute:

> A large amount of timber.

> A large number of students.

See also **less, fewer.**

analysis. The plural is **analyses**.

anyone, any one. Anyone is written as two words to give numerical emphasis; otherwise it is written as one word:

> Any one of us could do that.

> Anyone could do that.

anyways. Non-standard English: use *anyway*.

aqueduct. Note that the fourth letter in this word is **e**. Some similar words use an **a**, as in **aquamarine**, or an **i**, as in **aquifer**.

as, because. As is a weaker conjunction than **because** and may be confused with *when*:

> ✗ As I was working, I ate at my desk.

> ✓ Because I was working, I ate at my desk.

> ✗ He arrived as I was leaving.

> ✓ He arrived when I was leaving.

as to. A common feature of bureaucratese. Replace it with a single-word preposition such as *about* or *on*:

> ✗ They were concerned as to the range of disagreement.

> ✓ They were concerned about the range of disagreement.

> ✗ They recorded his comments as to the treaty.

> ✓ They recorded his comments on the treaty.

bad, badly. Bad is an adjective meaning *not good*:

> The meat tastes bad.

> He felt bad about forgetting the dinner party.

Badly is an adverb meaning *not well;* when used with the verbs **want** or **need**, it means *very much*:

> She thought he played the villain's part badly.

> I badly need a new suit.

beside, besides. Beside is a preposition meaning *next to*:

> She worked beside her assistant.

Besides has two uses: as a preposition it means *in addition to*; as a conjunctive adverb it means *moreover*:

> Besides recommending the changes, the consultants are implementing them.

> Besides, it was hot and we wanted to rest.

between. See **among**.

bite, byte. To **bite** is to *cut into* with the teeth; a **byte** is a single unit of information stored in a computer.

bring, take. One **brings** something to a closer place and **takes** it to a farther one:

> Take it with you when you go.

> Next time you come to visit, bring your friend along.

can, may. Can means to *be able*; **may** means to *have permission*:

> Can you fix the lock?

> May I have another piece of cake, please?

In speech, **can** is used to cover both meanings: in formal writing, however, you should observe the distinction.

can't hardly. A faulty combination of the phrases **can't** and **can hardly**. Use one or the other of them instead:

> He can't swim.

> He can hardly swim.

capital, capitol. As a noun **capital** may refer to a seat of government, the top of a pillar, an upper-case letter, or accumulated wealth. **Capitol** refers only to a specific American—or ancient Roman—legislative building.

circumlocution. A roundabout or circuitous expression: e.g., *in the family way* for *pregnant*; *at this point in time* for *now*.

cite, sight, site. To **cite** something is to *quote* or *mention* it as an example or authority; **sight** can be used in many ways, all of which relate to the ability to *see*; **site** refers to a specific *location*, a particular place at which something is located.

clayey. Since *clay* ends in a y, add *ey*—and not just *y*—to make it into an adjective:

The soils on his farm are very clayey.

climatic, climactic. Similar-sounding adjectives with entirely different meanings. **Climatic** refers to the *climate* of an area; **climactic** refers to the *climax* of something (for instance, the climactic scene in a play).

complement, compliment. The verb to **complement** means to *complete;* to **compliment** means *to praise.*

His engineering skill complements the skills of the designers.

I complimented her on her outstanding report.

compose, comprise. Both words mean *to constitute* or *make up*, but **compose** is preferred. **Comprise** is correctly used to mean *consist of*, or *be composed of*. Using **comprise** in the passive ("is comprised of")—as you might be tempted to do in the second example below—is usually frowned on in formal writing:

These students compose the group which will go overseas.

Base camp comprised a sleeping tent, a dining tent, and a latrine.

continual, continuous. Continual means *repeated over a period of time;* **continuous** means *constant* or *without interruption:*

The strikes caused continual delays in building the road.

In August, it rained continuously for five days.

could of. Incorrect, as are **might of, should of**, and **would of**. Replace **of** with have.

   ✗   He <u>could of</u> done it.

   ✓   He <u>could have</u> done it.

   ✓   They <u>might have</u> been there.

   ✓   I <u>should have</u> known.

   ✓   We <u>would have</u> left earlier.

council, counsel. Council is a noun meaning an *advisory* or *deliberative assembly*. **Counsel** as a noun means *advice* or *lawyer;* as a verb it means to *give advice.*

> The college <u>council</u> meets on Tuesday.

> We respect his <u>counsel</u>, since he's seldom wrong.

> As a camp <u>counsellor</u>, you may need to <u>counsel</u> parents as well as children.

criterion, criteria. A **criterion** is a standard for judging something. **Criteria** is the plural of **criterion** and thus requires a plural verb:

> These <u>are</u> my criteria for selecting the paintings.

data. The plural of **datum**. The set of information, usually in numerical form, that is used for analysis as the basis for a study. Informally, **data** is often used as a singular noun, but in formal contexts it should be treated as a plural:

> <u>These</u> data <u>were</u> gathered in an unsystematic fashion. Therefore they are inconclusive.

> John reported that his data <u>are</u> exactly what he needs.

deduce, deduct. To **deduce** something is to *work it out by reasoning;* to **deduct** means to *subtract* or *take away* from something. The noun form of both words is **deduction.**

defence, defense. Both spellings are correct: **defence** is common in Britain, **defense** in the United States.

delusion, illusion. A delusion is a belief or perception that is distorted; an **illusion** is a false belief:

> Hitler had delusions of grandeur.

> He thought he saw a sea monster, but it was an illusion.

dependent, dependant. Dependent is an adjective meaning *contingent on* or *subject to*; **dependant** is a noun.

> Andrew's graduation is dependent upon his passing algebra.

> Andrew is a dependant of his father.

depletion, depression, deprivation. Depletion means *using up* (as with a non-renewable resource) or *reducing* the amount or quality of a (renewable) resource to the point where it cannot easily recover. **Depression** literally means *pressing down*. Thus in economic terms it refers to a marked downturn or trough in the business cycle, which can cause extreme economic hardship; in physical terms it refers to a marked trough, as in a *cyclonic depression*, with associated violent weather conditions. **Deprivation** is the denial of access to basic human requirements.

device, devise. The word ending in **-ice** is the noun; that ending in **-ise** is the verb.

> A ground penetrating radar (GPR) is a useful device.

> She is working to devise a model of eco-industrial development.

different than. Incorrect. Use either **different from** (American usage) or **different to** (British).

diminish, minimize. To **diminish** means to *make* or *become smaller;* to **minimize** is to *reduce* something to the smallest possible amount or size.

disinterested, uninterested. Disinterested implies impartiality or neutrality; **uninterested** implies a lack of interest:

> As a disinterested observer, he was in a good position to judge the issue fairly.

> Uninterested in the proceedings, he yawned repeatedly.

due to. Although increasingly used to mean *because of*, **due** is an adjective and therefore needs to modify something:

> ✗   Due to his impatience, we lost the contract. [Due is dangling.]

> ✓   The loss was due to his impatience.

e.g., i.e. E.g. means *for example*; **i.e.** means *that is*. The two are incorrectly used interchangeably. See **Latin abbreviations** (Chapter 18).

entomology, etymology. Entomology is the study of insects; **etymology** is the study of the derivation and history of words.

exceptional, exceptionable. Exceptional means *unusual* or *outstanding*, whereas **exceptionable** means *open to objection* and it is generally used in negative contexts.

> His accomplishments are exceptional.

> There is nothing exceptionable in his behaviour.

farther, further. Farther refers to distance, **further** to extent; more:

> He paddled farther than his friends.

> He explained the plan further.

focus (noun). The plural may be either **focuses** (also spelled **focusses**) or **foci**.

good, well. Good is an adjective that modifies a noun; **well** is an adverb that modifies a verb.

> He is a good rugby player.

> The experiment went well.

hanged, hung. Hanged means *executed by hanging.* **Hung** means *suspended* or *clung to*:

> He was hanged at dawn for the murder.

> He hung the picture.

> He hung on to the boat when it capsized.

hereditary, heredity. Heredity is a noun; **hereditary** is an adjective. **Heredity** is the biological process whereby characteristics are passed from one generation to the next; **hereditary** describes those characteristics.

> Heredity has determined that you have brown hair.

> Your short legs must be hereditary.

hopefully. Use **hopefully** as an adverb meaning *full of hope*:

> She scanned the horizon hopefully, waiting for her friend's ship to appear.

In formal writing, using **hopefully** to mean *I hope* is still frowned upon, although increasingly common; it's better to use *I hope*:

> ✗   Hopefully the experiment will go off without a hitch.

> ✓   I hope the experiment will go off without a hitch.

i.e. *Not* the same as **e.g.**! See **e.g.**

illusion. See **allusion; delusion.**

incite, insight. Incite is a verb meaning to *stir up*; **insight** is a noun meaning (often sudden) understanding.

infer, imply. To **infer** means to *deduce* or *conclude by reasoning*. It is often confused with **imply**, which means to *suggest* or *insinuate*.

> We can infer from the large population density that there is a large demand for services.

> The large population density implies that there is a high demand for services.

inflammable, flammable, non-flammable. Despite its **in-** prefix, **inflammable** means the same as **flammable**: it describes things that are *easily* set on fire; **non-flammable** means the opposite. To prevent any possibility of confusion, it's best to avoid **inflammable** altogether.

irregardless. Redundant; use *regardless*.

italics. Slanting type used for emphasis; sometimes replaced in typescript by underlining.

its, it's. Its is a possessive pronoun; **it's** is a contraction of *it is*. Many people mistakenly put an apostrophe in **its** in order to show possession.

✗ The cub wanted <u>it's</u> mother.

✓ The cub wanted <u>its</u> mother.

✓ <u>It's</u> time to leave.

less, fewer. Less is used when units are *not* discrete and *not* absolute ("less woodland"). **Fewer** is used when the units *are* discrete and absolute ("fewer people").

lie, lay. To **lie** means to *assume a horizontal position;* to **lay** means to *put down.* The changes of tense often cause confusion:

Present	*Past*	*Past participle*
lie	lay	lain
lay	laid	laid

like, as. Like is a preposition, but it is often wrongly used as a conjunction. To join two independent clauses, use the conjunction **as:**

✗ I want to progress <u>like</u> you have this year.

✓ I want to progress <u>as</u> you have this year.

✓ Prof. Dodd is <u>like</u> my old school principal.

might of. See **could of.**

minimize. See **diminish.**

mitigate, militate. To **mitigate** means to *reduce the severity* of something; to **militate** against something means to *oppose* it.

myself, me. Myself is an intensifier of, not a substitute for, *I* or *me:*

✗ He gave it to John and <u>myself</u>.

✓ He gave it to John and <u>me</u>.

✗ Jane and <u>myself</u> are invited.

✓ Jane and <u>I</u> are invited.

✓ <u>Myself</u>, <u>I</u> would prefer a swivel chair.

nor, or. Use **nor** with **neither** and **or** by itself or with **either**:

He is <u>neither</u> overworked <u>nor</u> underfed.

The plant is <u>either</u> diseased <u>or</u> dried out.

off of. Remove the unnecessary **of**:

✗ The fence kept the children <u>off of</u> the premises.

✓ The fence kept the children <u>off</u> the premises.

phenomenon. A singular noun: the plural is **phenomena**.

plaintiff, plaintive. A **plaintiff** is a person who brings a case against someone else to court; **plaintive** is an adjective meaning sorrowful.

populace, populous. Populace is a noun meaning the *people* of a place; **populous** is an adjective meaning *thickly inhabited*.

The <u>populace</u> of Hilltop village is not well educated.

With so many people in such a small area, Hilltop village is a <u>populous</u> place.

practice, practise. Practice can be a noun or an adjective; **practise** is always a verb. Note, however, that in the U.S. and sometimes in Canada, the spelling of the verb is **practice**:

The soccer players need <u>practice</u>. (noun)

That was a <u>practice</u> game. (adjective)

The players need to <u>practise</u> (or <u>practice</u>) their skills. (verb)

precede, proceed. To **precede** is to *go before* (earlier) or *in front of* others; to **proceed** is to *go on* or *ahead*.

The faculty will <u>precede</u> the students into the hall.

The medal winners will <u>proceed</u> to the front of the hall.

prescribe, proscribe. These words are sometimes confused, although they have quite different meanings. **Prescribe** means *to advise the use of* or *impose authoritatively*. **Proscribe** means to *reject, denounce,* or *ban*:

> The professor <u>prescribed</u> the conditions under which the equipment could be used.

> The student government <u>proscribed</u> the publication of unsigned editorials in the newspaper.

principle, principal. **Principle** is a noun meaning a *general truth* or *law*; **principal** can be used as either a noun or an adjective, meaning *chief*.

rational, rationale. **Rational** is an adjective meaning *logical* or *able to reason*. **Rationale** is a noun meaning *explanation*:

> That was not a <u>rational</u> decision.

> The president sent around a memo with a <u>rationale</u> for his proposal.

real, really. **Real**, an adjective, means *true* or *genuine;* **really**, an adverb, means *actually, truly, very,* or *extremely*.

> The nugget was <u>real</u> gold.

> The nugget was <u>really</u> shiny and <u>really</u> worth a lot.

Scots, Scottish, Scotch. **Scots** and **Scottish** mean the same thing: *of Scotland* (e.g., people, accents, etc.). **Scotch** is used in various compound nouns, such as *Scotch mist, terrier,* and of course, *whiskey*. Never refer to a *Scotsman* or *Scotswoman* as a Scotchman or Scotchwoman.

sceptic, septic, septic tank. A **sceptic** is one who *doubts* or who *critically questions;* **septic** means *infected* with harmful micro-organisms that cause pus to form, as in a septic wound; a **septic tank** is a container into which sewage is conveyed and in which it remains until the activity of bacteria makes it liquid enough to drain away.

seasonable, seasonal. Seasonable means *usual* or *suitable for the season*; **seasonal** means *of, depending on,* or *varying with the season*:

> The temperature is seasonably high.

> The clothes you pack must take into account seasonal changes in the weather.

should of. See **could of.**

stratum. A singular noun. The plural is **strata.**

their, there. Their is the possessive form of the third person plural pronoun. **There** is usually an adverb, meaning *at that place* or *at that point*; sometimes it is used as an expletive (an introductory word in a sentence):

> They parked their bikes there.

> There is no point in arguing with you.

> There, look at the whales!

tortuous, torturous. The adjective **tortuous** means *full of twists and turns* or *circuitous.* **Torturous,** derived from *torture,* means *involving torture* or *excruciating*:

> To avoid heavy traffic, they took a tortuous route home.

> The band concert was a torturous experience for the audience.

towards. Non-standard English: use *toward*

translucent, transparent. A **translucent** substance permits light to pass through, but not enough for a person to see through it; a **transparent** substance permits light to pass unobstructed, so that objects can be seen clearly through it.

turbid, turgid. Turbid, with respect to a liquid or colour, means *muddy, not clear,* or (with respect to literary style) *confused.* **Turgid** means *swollen, inflated,* or *enlarged,* or (again with reference to literary style) *pompous* or *bombastic.*

unique. This word, which means *of which there is only one* or *unequalled,* is both overused and misused. Since there are no degrees of comparison—one thing cannot be more unique than another—expressions such as **very unique** are incorrect.

while. To avoid misreading, use **while** only when you mean *at the same time that.* Do not use *while* as a substitute for *although, whereas,* or *but:*

✗ While he's getting fair marks, he'd like to do better.

✓ I headed for home, while she decided to stay.

✓ He fell asleep while he was reading.

-wise. Never use **-wise** as a suffix to form new words when you mean *with regard to:*

✗ Sales-wise, the company did better last year.

✓ The company's sales increased last year.

would have, would of. When people are describing a hypothetical instance in the past, they often mistakenly use **would have** in both the condition (*if*) clause and the consequence clause: see p. 235. For **would of,** see **could of.**

your, you're. Your is a pronominal adjective used to show possession; **you're** is a contraction of *you are:*

Your sleeping bag is warmer than mine.

You're likely to miss your train.

CHAPTER 18

Definitions

abbreviations. Shortened forms of words. There are two types: (a) true abbreviations, in which only the first few letters of the full word are used: e.g., *vol.* for *volume, cont.* for *continued*; and (b) suspensions, which include the last letter of the full word: e.g., *figs.* for *figures, Dr.* for *doctor.* This book follows the most common North American style, which uses a period after suspensions as well as contractions. However, some writers omit the period after suspensions, since these include the final letter. See also **acronyms, contractions, Latin abbreviations.** Common abbreviations include:

abbr.	abbreviation, abbreviate
abr.	abridged
acad.	academy
adapt.	adapted from (or by), adaptation
Amer.	American
assn.	association
attrib.	attributed to
bibliog.	bibliography, bibliographer, bibliographic
biog.	biography, biographer, biographic
bull.	bulletin
(c)	copyright
cf.	compare
colloq.	colloquial
comp.	compiler, compiled by
conf.	conference
diss.	dissertation
div.	division
doc.	document
ed.	editor, edited by, edition
ex.	example
fwd.	foreword
geog.	geography
geol.	geology

illus.	illustration
inst.	institute, institution
intr., introd.	introduction
jour.	journal
misc.	miscellaneous
ms., mss.	manuscript, manuscripts
n., nn.	note, notes (e.g., "64, nn. 8–19": "page 64, notes 8 to 19")
n.d.	no date
no.	number
n.p.	no place of publication cited, no publisher
n.pag.	no pagination
n.s.	new series (of a journal or monograph series)
orig.	original
o.s.	old (original) series (of a journal or monograph series)
proc.	proceedings
publ.	publisher, publication, published by
qtd.	quoted
rept.	report
rev.	review, revised, revision
ser.	series (e.g., of books on a given theme)
supp.	supplement
trans., tr.	translator, translation, translated by
UP	University Press (e.g., Oxford UP)
var.	variant
vers.	version

abstract. An **abstract** is a *brief statement* summarizing the main points of a work. Abstracts usually are required for formal scientific reports or papers, and upper-level undergraduate and graduate theses. The adjective **abstract** means *theoretical* rather than concrete.

abstract language. Language that deals with theoretical, intangible concepts or details, e.g., justice, goodness, truth. (Compare **concrete language**.)

accessibility. The relative opportunity for interaction and contact.

acculturation. The process of adopting a culture; generally used in reference to an immigrant group's acceptance of the culture of the host society over a period of time. See also **assimilation** and **integration**.

acronyms, initialisms. Words formed by joining together the first letters from a set of words. Periods are generally not used in acronyms: e.g., *NATO* from *North Atlantic Treaty Organization*. Names in which each letter is pronounced separately (e.g., *CBC:* Canadian Broadcasting Corporation) are called **initialisms**. A few acronyms have been accepted as words in their own right: e.g., *laser*, from *light amplification* (by) *stimulated emission* (of) *radiation*.

adaptation. The process whereby a living organism changes in order to adjust to and thrive in a particular environment.

adjective. A word that modifies or describes a noun or pronoun; hence a kind of noun marker: e.g., *red, beautiful, solemn*. An **adjectival phrase** or **adjectival clause** is a group of words modifying a noun or pronoun.

adverb. A word that modifies or qualifies a verb, adjective, or adverb, often answering a question such as *how? why? when?* or *where?*: e.g., *slowly, fortunately, early, abroad*. An **adverbial phrase** or **adverbial clause** is a group of words modifying a verb, adjective, or adverb: e.g., *by force, in revenge*. See also **conjunctive adverb**.

agreement. Consistency in tense, number, or person between related parts of a sentence: e.g., between subject and verb, or noun and related pronoun.

allegory. A story in which the characters and events stand for someone or something else.

alliteration. The repetition of the first letter of one word in the word or words that follow, a stylistic technique that, used sparingly, can have a good effect: e.g., *rushing river*.

ambiguity. Vague or equivocal language; meaning that can be taken two ways.

ampersand (&). The character &, meaning *and*.

analogy. An argument in which one thing is likened to another; often the parallel is between something that is little understood and something more familiar: e.g., *Insulation is to a house as clothes are to the human body*.

antecedent (referent). The noun for which a pronoun stands.

antonyms. Pairs of words with opposite meanings: e.g., *quick* and *slow; hot* and *cold*. See **synonyms**.

appendix. An additional section following the main body of a work, containing material that contributes but is not essential to the completeness of the work.

appositive. A word or phrase that identifies a preceding noun or pronoun: e.g., *Mrs. Jones,* **my aunt,** *is sick.* The second phrase is said to be in *apposition* to the first.

appropriate technology. Generally, technology that is appropriate to a situation. A more specific meaning has developed in the context of the developing countries of the **third world**, referring to the application of technological concepts and innovations that (a) are either indigenous or not too far removed (in time or space) from the comprehension and skills of the society using them, and (b) produce the desired results without unfairly exploiting the people or their resources.

areal differentiation. The differences in and between various areas on the surface of the earth from the perspectives of the *interrelations* between phenomena, the *differential character* of the phenomena and the complexes they form, and their *areal* expression.

article. A word that precedes a noun and shows whether the noun is definite or indefinite; a kind of determiner or noun-maker. **Indefinite article**: *a (an)*. **Definite article**: *the*.

assertion. A positive statement or claim: e.g., *These findings are irrefutable.*

assimilation. In *sociology*, the process whereby a minority becomes increasingly like the dominant society; **assimilation** implies that the minority will lose its cultural distinctiveness. See **integration**. In *science*, the term refers to conversion into a similar substance; e.g., the conversion by an animal or plant of extraneous material into fluids and tissues identical with its own.

assonance. Repetition of the same vowel sound in a series of words for a special effect: e.g., *He drove home so slowly.*

auxiliary. A verb used in combination with another verb to create a verb phrase; a helping verb used to create certain tenses and emphases: e.g., *could, do, may, will, have.*

average, mean, median. An **average** is obtained by adding several quantities together and dividing the total by the number of quantities. The **mean** is between the first and last of an arithmetical or geometrical progression; thus 2 and 8 have arithmetic mean 5, and geometric mean 4. **Mean** also refers to the midpoint between two extremes (e.g., **mean sea level** is the halfway point between high and low water). The **median** is the middle number in a given set of numbers (e.g., the median of 3, 5, 7, 12, 65 is 7, but the median of 3, 5, 7, 10 is 6—halfway between the two middle numbers).

B.C., B.C.E. B.C. means *before Christ;* **B.C.E.** means *before the Common Era.*

bibliography. (a) A list of works referred to or found useful in the preparation of an essay or report; (b) a reference book listing works available on a particular subject.

bioenergetics. The study of energy production and flow within and between organisms.

biofuel. Abbreviation of **biomass fuel,** which is fuel derived from renewable organic sources.

biography, autobiography. A **biography** is a written account of another person's life that generally adheres to facts; an **autobiography** is a biography of oneself.

biomass. The total weight of organic substances found in a defined area; a renewable source of energy.

biome. See **ecosystem.**

boundary, frontier. A **boundary** is a *line* as the limit of any territorial unit, whether physical, social, political, or economic. A boundary that is *delimited* is merely agreed to on paper; one that is *demarcated* is marked on the ground also. A **boundary** also can be a dividing line between contrasting cultural and political systems, with competing constructions of socio-spatial consciousness that in turn are related to core notions of nationalism.[1] A **frontier** is an *area* at the margins of a territorial unit into which expansion can take place. See **territory.**

carrying capacity. Originally an ecological term used to refer to the maximum number of plants that can be sustained in a particular area. Now often used to refer to the maximum number of users (animals and people) that can be sustained by the land resources in a particular area (such as a national park).

case. The inflected form of pronouns (see **inflection**). **Subjective case**: *I, we, he, she, it, the.* **Objective case**: *me, us, him, her, it, them.* **Possessive case**: *my, our, his, her, its, their.*

centrifugal, centripetal. Centrifugal means *moving away from* the centre or axis; **centripetal** means *moving toward* the centre or axis. The former is used literally in reference to a *centrifuge*. In addition, human geographers use both terms to refer to the counteracting processes that produce certain conditions within states, or that produce certain patterns of population and land use at more local levels, as in cities.

circulation. Used in human geography to refer to the myriad ways in which **spatial** movement (of ideas, people, and things) can occur. The term implies overcoming difficulties caused by distance and the geographical nature of the intervening space.

circumlocution. A roundabout or circuitous expression: e.g., *in the family way* for *pregnant*; *at this point in time* for *now*.

classification, regionalization. Classification is the differentiation of things (objects, people, places, ideas, etc.) into set groups according to predetermined criteria; **regionalization** is a particular type of classification, in which places or areas and their contents are grouped according to selected criteria.

clause. A group of words containing a subject and predicate. An **independent clause** can stand by itself as a complete sentence: e.g., *I bought a hamburger*. A **subordinate** or **dependent clause** cannot stand by itself but must be connected to another clause: e.g., **Since I was hungry**, *I bought* a hamburger.

cliché. A trite or well-worn expression that has lost its impact through overuse: e.g., *slept like a log, sunny disposition, tried and true.*

climax. Point of greatest intensity; culmination; the final, self-sustaining, stable plant community that is formed after a series of developments, with the changes coming more slowly as the climax stage is approached. See also **equilibrium**.

collective noun. A noun that is singular in form but refers to a group: e.g., *family, team, jury*. It may take either a singular or a plural verb, depending on whether it refers to individual members or to the group as a whole.

colloquialisms. Expressions that are used in speech but generally not in writing, unless the writing style is informal or the phrase is used in written dialogue, as in a novel or play: e.g., *Let's grab a pizza.*

comma splice. See **run-on sentence**.

complement. A completing word or phrase that usually follows a linking verb to form a **subjective complement**: e.g., (1) *He is* **my father**. (2) *That cigar smells* **terrible**. If the complement is an adjective it is sometimes called a **predicate adjective.** An **objective complement** completes the direct object rather than the subject: e.g., *We found him* **honest and trustworthy**.

complementarity. Used in economic geography to refer to the relation between two regions wherein one region produces goods that the other region does not have but wants. **Complementarity** is the basis for **spatial interaction** (see **spatial**).

complex sentence. A sentence containing a dependent clause as well as an independent one: e.g., *I bought the ring, although it was expensive.*

compound sentence. A sentence containing two or more independent clauses: e.g., *I saw the oil spill and I reported it.* A sentence is called **compound-complex** if it contains a dependent clause as well as two independent ones: e.g., *When the fog lifted, I saw the oil spill and I reported it.*

conclusion. The part of an essay in which the findings are pulled together or the implications are revealed so that the reader has a sense of closure or completion.

concrete language. Specific language, giving particular details (often details of sense): e.g., *red corduroy dress, three long-stemmed roses.* See **abstract language**.

conditional. The conditional form of the verb expresses what may happen if a particular condition applies: e.g., **If it rains**, *the field trip will be cancelled.*

conjunction. An uninflected word used to link words, phrases, or clauses. A **coordinating conjunction** (e.g., *and, or, but, for, yet*) links two equal parts of a sentence. A **subordinating conjunction**, placed at the beginning of a subordinate clause, shows the logical dependence of that clause on another: e.g., (1) **Although** *I am poor, I am happy.* (2) **While** *others slept, he studied.* **Correlative conjunctions** are pairs of coordinating conjunctions (see **correlatives**).

conjunctive adverb. A type of adverb that shows the logical relation between the phrase or clause that it modifies and a preceding one: e.g., (1) *I sent the letter; it never arrived,* **however**. (2) *The battery died;* **therefore** *the car wouldn't start.*

connotation. Associative meaning; the range of suggestion called up by a certain word. Apparent synonyms, such as *poor* and *underprivileged,* may have different connotations. See **denotation**.

consistency. Use of the same style (of spelling, punctuation, documentation) throughout a given piece of work.

context. The framework provided by others' research, within which a given study is situated; the text surrounding a particular passage that helps to establish its meaning.

contractions. Words formed by combining and shortening two words: e.g., *isn't, can't, we're.* Contractions are permissable in informal writing.

coordinate construction. See **correlatives**.

coordinating conjunction. See **conjunction**.

copula verb. See **linking verb**.

correlatives (coordinates). Pairs of coordinating conjunctions: e.g., *either/or, neither/nor, not only/but.*

culture. Derived from the Latin *colere,* to cultivate. Used in biology and medicine with reference to "tissue culture": the propagation of micro-organisms (plant or animal) or of living tissue cells in special media that are conducive to their growth. **Culture** also is used widely in the social sciences, where it generally refers to a wide range of characteristics that help to define a population as a distinctive unit. *Elements* of **culture** are universally shared categories (including language, religion, politics, and family), but the particular *traits* or characteristics differ from group to group (hence a particular language, particular ways of doing things, particular social structures, etc.). **Culture** is *not* synonymous with **race**. See also the next entry.

culture area, culture history, cultural ecology, cultural landscape. Terms used in cultural geography, anthropology, and, selectively, other disciplines. **Culture area** refers to the distributional aspect of culture, with emphasis on regions; **culture history** refers to the historical dimension, especially from the perspective of origins and dispersals; **cultural ecology** is the study of the relationships between a group and its physical surroundings; **cultural landscape** is the physical environment as moulded and modified by human action. The study of any one of these themes necessarily involves consideration of the others.

cycle, recycle. A **cycle** is a series of events that recurs regularly, returning to a similar starting position or state after a particular period of time. There are many cycles, including the birth, death, decay, and rebirth of plants; diurnal (daily) and seasonal cycles; the hydrologic cycle; the flows of elements and nutrients in the biosphere; the lunar cycle; and the cycle of agricultural livelihood patterns. To **recycle** means to *reclaim* materials for further use that would otherwise be wasted; the recycling process is not a true cycle.

dangling modifier. A modifying word or phrase (often a participial phrase) that is not connected grammatically to any part of the sentence: e.g., **Walking to school**, *the street was slippery.*

demonstrative pronoun. A pronoun that points out something: e.g., (1) **This** *is his reason.* (2) **That** *looks like my lost earring.* When used to modify a noun or pronoun, a demonstrative pronoun becomes a kind of **pronominal adjective**: e.g., *this hat, those people.*

denotation. The literal or dictionary meaning of a word. See **connotation**.

dependent clause. See **clause**.

diction. The choice of words with regard to their tone, degree of formality, or register. Formal diction is the language of orations and serious essays. The informal diction of everyday speech or conversational writing can, at its extreme, become slang.

direct/indirect speech (discourse). **Direct speech** (or **discourse**) consists of the actual words a person spoke or wrote. In writing, these words are enclosed in quotation marks: *The politician said, "Environmental issues are of prime importance to this government."* **Indirect speech** (or **discourse**) gives the meaning of the speech rather than the actual words. In writing, indirect discourse is not enclosed in quotation marks: *He said that environmental issues were very important to the government.*

discourse. Talk, either oral or written. **Discourse** may be **direct** or **indirect**; see the preceding entry.

eco-. A prefix that has gained favour in the past four decades to qualify things related—directly or even vaguely—to environmental issues. Examples include *ecoaccident, ecoactivist, ecocrisis, ecoethics, ecospeak, ecotourism,* and *econut.*

ecology. The study of relationships among living organisms with due concern for their interactions with the physical environment.

ecosystem. An areally distinct community of interdependent organisms together with their environment. For instance, a pond inhabited by a particular species of frog may function as an ecosystem that is distinct from the fauna functioning as an ecosystem in an adjacent environment. An especially large ecosystem, such as the southern African savannah grasslands, may be referred to as a **biome.**

editing. The process of checking what you have written for problems with spelling, grammar, expression, and content; an essential step in the writing process.

ellipsis marks. Spaced dots indicating an omission from a quoted passage. (See p. 250.)

emigrate, immigrate, migrate, migratory, migration. To **emigrate** is to *leave* one country for another; to **immigrate** is to *come to* one country from another. Both terms imply a person's permanent relocation from one country to another. Birds and some mammals **migrate** by changing their place of living from one season to another; **migratory** birds are the ones that **migrate.** In each case, the act of moving is **migration.**

enclave, exclave. In political geography, both words refer to the same thing— a small territory belonging to one state that is surrounded by territory of another state (or states). However, the two terms reflect two different points of view: **enclave,** that of the surrounding territory; **exclave,** that of the home state. In economic geography, **enclave** refers to a clustering of modern industry and associated facilities in an otherwise underdeveloped country.

endnote. A footnote or citation placed at the end of an essay or report.

environment. Often misspelled by students (e.g., "enviroment," "envirenment,". "envivionment"). A useful, often abused word that means different things to different people. Disciplines, too, define **environment** in different ways. For example, the term means something quite different to (1) a nurse concerned with an individual's "environment"—meaning the total set of stressors that impinge upon that person's system, together with the resources available to cope with those stressors; (2) a soil scientist concerned with local chemical and physical factors in a surface or near-surface environment; (3) a biologist concerned with the ecological environs of a lake shore; (4) an urban geographer concerned with the physical environment and, in particular, the built environment in and on which people have an impact; (5) a planner concerned with the built environment as the place that can be purposefully and effectively changed to reflect some planned outcome; (6) a social or cultural geographer concerned with how people react to their surroundings and attribute meaning to those surroundings; and so on. Thus, **environment** can refer to the physical characteristics of places or the cultural settings of places and people, and may be qualified as necessary: *behavioural* environment, *phenomenal* environment, *social* environment, *economic* environment, *urban* environment, *religious* environment, and so on. In the broadest sense, **environment** also is used to mean **nature**.

environmental sciences. The many disciplines that study the biological, geographical, and physical world, with varying concern for the functioning of organisms, the physical environment, and the interaction between the two. The disciplines and areas of interdisciplinary focus include agriculture, architecture, biology, botany, chemistry, earth and atmospheric science, ecology, environmental economics, environmental engineering, environmental monitoring and analysis, environmental protection, forestry, **geography**, geology, horticulture, land economics, land resources, natural resources management, physics, toxicology, water resources, and zoology. Although common core requirements for students in the environmental sciences may suggest close links between the various disciplines, too often disciplinary parochialism and differences in the scales of research foci serve to discourage collaboration.

environmental studies. Interdisciplinary programs that seek to study environmental issues in an integrative manner, drawing from the biological, geographical, physical, and social sciences, and the humanities.

equi-. The prefix meaning *equal*; hence *equiangular, equidistant, equilateral, equilibrate,* **equilibrium**.

equilibrium. Balance; in ecological terms, refers to a system in which inputs and outputs are finely, though not statically, balanced; hence *dynamic equilibrium*.

essay. A literary composition on any subject. Some essays are descriptive or narrative, but in an academic setting most are expository (explanatory) or argumentative.

euphemism. A word or phrase used to avoid some other word or phrase that might be considered offensive in some way: e.g., instead of saying *I have to pee*, someone might say *I have to see a man about a dog*.

expletive. A grammatically meaningless exclamation or phrase. The most common expletives are the sentence beginnings *It is* and *There is (are)*.

exploratory writing. The informal writing done to help generate ideas before formal planning begins.

figurative. The opposite of **literal**. **Figurative meaning** is abstract or imaginative; words used **figuratively** are given an abstract or imaginative meaning rather than their literal or face-value meaning.

figure of speech. A **figurative** expression used to create a special effect. See **hyperbole**, **imagery**, **metaphor**, and **simile**.

first world, second world, third world, fourth world. The first three phrases refer to categories of states within the world economy. The **first world** consists of the advanced capitalist, industrialized countries of the West (notably the United States, Canada, Western Europe, and Japan). **Second world** refers to the former state-socialist states (of the U.S.S.R. and Eastern Europe)—although the term has lost its usefulness since the early 1990s, with the development of new external diplomatic, trade, and military alignments among some of these states. **Third world** refers to "underdeveloped" states, especially in Africa, Asia, and Latin America, not in the first or second worlds. **Fourth world** initially referred to indigenous peoples. However, the term also has been used to refer to people in the Pacific islands; a people within a state who have a distinctive identity and desire self-determination; and a subset of states in the **third world** that are the poorest of the poor. Among the other terms that have gained currency as descriptors of the major economic and technological divisions between states are *developed* versus *developing* and *North* ("rich") versus *South* ("poor").

footnote. A citation placed at the bottom of a page or the end of the composition. See **endnote**.

foreword, preface. A brief section at the beginning of a book in which the author or editor explains his or her intentions and how the book came to be written; often includes acknowledgements, although these may be presented separately.

frontier. See **boundary**

fused sentence. See **run-on sentence**.

Gaia. A recently developed concept that refers to the living "being" of the biosphere.

general language. Language lacking specific details; abstract language.

geography, geographers. Widely used terms with numerous meanings. Geography is from the Greek *geo,* the earth, and *graphein,* to write; **geographers** "do" geography. Geography is commonly defined as the study of the earth's surface and the people who inhabit it, or "the study of the earth as the home of people."[2] The study of locations and place names is to geography as the skeleton is to medicine: simply the basic framework. **Geographers** focus on **spatial** distributions, patterns, processes, and relationships between human societies, the physical **environment, nature,** the **cultural landscape, locations, circulation,** and **regional** structures. **Geographers** who stress physical properties and processes are more aligned with the physical sciences; those who stress societal organization and human behaviour set within particular places are more aligned with the social sciences and, in some cases, the humanities. Bridging the "pure" or "hard" sciences, the social sciences, and, at times, the arts, geography is a correlative discipline.[3] Geographers are uncomfortable with single-perspective explanations, because they tend to see patterns and interrelationships that students in other disciplines may miss. The **scale** of approach varies from the personal to small groups to large groups, from the laboratory to small areas to larger regions to the world itself. All geographers share the goal of making "sense of that vast panorama of fact and fiction, pattern and event, on the surface of the earth."[4] Areas of study range from integrative regional geography through many systematic specializations, including physical, political, cultural, economic, social, feminist, historical, urban, climatology, geomorphology, soils, fluvial, environmental perception, tourism, sport, and more.[5] Much of the work is of an applied nature, including electoral redistricting, natural disaster recovery, river sedimentation, resource analysis and management, locational analysis for siting commercial and industrial functions, and working with people with disabilities. Some work highlights public policy issues, whether domestic or international, including global warming and terrorism.[6] Both qualitative and quantitative techniques are used, the former by cultural and humanistic geographers, the latter by physical and economic geographers. Also used are air photo interpretation, traditional cartographic methods, computer-assisted cartographic techniques, and, relatedly, GIS (Geographic Information Systems) techniques.

gerund. A verbal (part-verb) or verb form that functions as a noun and is marked by an -*ing* ending: e.g., **Swimming** *can help you become fit.*

globalization. The increasingly important form of capitalism characterized by the increased mobility of capital and the rise of transnational corporations, facilitated by the rapid evolution of worldwide digital communications technology. See **restructuring**.

grammar. The study of the forms and relations of words, and of the rules governing their use in speech and writing.

habitat. The dwelling area of a species or community; in biology and biogeography, **habitat** generally refers to a specific type of habitat and its location, such as a forest floor, a wetland, or a seashore; in human geography **habitat** can refer to physical structures (e.g., farm dwellings) and their local environmental settings.

headings. The titles given to chapters in a work, or sections and subsections, indicating to readers the topics that are about to be discussed.

holism, holistic. In philosophy, **holism** is the theory that the whole is greater than the sum of its parts. Biologists use the term in the context of characteristics of organisms that are not functions of their individual components. Geographers attempt to be **holistic** in their approach, whether to the analysis of regions or to the interrelated physical and human dimensions of problems.

homonyms. Sets of words with different meanings that have the same sound or spelling: e.g., *site, sight, cite; (tent) pole, (magnetic) pole.*

hyperbole. Exaggeration for emphasis: e.g., *You've been told a million times to tie your shoelaces!*

hypothesis. A supposition or trial proposition made as a starting point for further investigation. The plural is *hypotheses*.

hypothetical instance. A supposed occurrence; often shown by a clause beginning with *if*.

imagery. Language that creates pictures in the reader's mind; see **metaphor** and **simile**. Cultural geographers writing about landscape often examine imagery.

impersonal writing. Writing that does not express the character of the writer. Scientific writing is impersonal.

independent clause. See **clause**.

indirect discourse. See **direct/indirect speech**.

infinitive. A type of verbal (or verb form) not connected to any grammatical **subject**: e.g., *to ask*. The **base infinitive** omits the *to*: e.g., *ask*.

inflection. The change in the form of a word to indicate number, person, case, tense, or degree.

informal language. Language, written or spoken, with many **colloquial** words and phrases. Not used in essays—unless for a creative writing class—or other formal writing.

initialisms. See **acronyms**.

integration. Used in the social sciences, including human geography, to refer to the process or situation in which subgroups within a society mix with the broader group while retaining their distinctive identity and cultural separateness; **integration** contrasts with **assimilation**. **Integration** also is used to refer to economic unions resulting from trade agreements (e.g., those between Canada, the U.S., and Mexico, or Australia and New Zealand), and, more broadly, instances of overall social, political, and economic union (such as in the European Union).

intensifier (qualifier). A word that modifies and adds emphasis to another word or phrase: e.g., **very** *tired*, **quite** *happy*, *I* **myself**.

interjection. A remark or exclamation interposed or thrown into a speech, usually accompanied by an exclamation mark: e.g., *Oh dear! Alas!*

interrogative sentence. A sentence that asks a question: e.g., *What is the time?*

intransitive verb. A verb that does not take a direct object: e.g., *fall, sleep, talk*.

introduction. A section at the start of an essay that tells the reader what is going to be discussed and why; an extremely important element of any essay.

italics. Slanting type used for emphasis, sometimes replaced in typescript by underlining.

jargon. Technical terms used unnecessarily or in inappropriate places: e.g., *peer-group interaction* for *friendship*.

landscape. A term used widely to refer to the characteristics of the earth's surface resulting from the interplay of myriad physical and human processes. Physical geography focuses on the processes (fluvial, wind-blown, etc.) that shape the landscape. Human geography considers landscape from many perspectives: as *scenery* (with emphasis on aesthetic aspects), as **nature**, as **habitat**, as *artifact* (i.e., the product of human activity), as **place** (wherein people interact), and as **territory**. The term **landscape** is sometimes used as shorthand for **cultural landscape**. There is an extensive interdisciplinary literature on landscapes, stressing thematic concerns (e.g., woodlands, agricultural clearances, drainage systems, rural settlements, landscapes of leisure, landscapes of political and ideological conflict) at world, regional, and more local scales. Geographers learn to *read* landscapes as if they were paintings or texts, and seek to reveal and interpret symbolic meanings that are imbedded in them.[7] The value of historical landscapes as *heritage* has proven to be a contentious subject.[8]

Latin abbreviations and terms. Many common abbreviations are derived from Latin words and phrases. Note where periods appear in these examples: *cf.* (from *confer*, "compare"); *e.g.* (from *exempli gratia*, "for example"); *etc.* (from *et cetera*, "and the rest"); *i.e.* (from *id est*, "that is"). One Latin term that is not abbreviated is *idem*, meaning "the same author" or "the same place." Today these terms usually are not italicized (but see p. 244 on *sic*).

linking verb (copula verb). The verb *to be* used to join subject to complement: e.g., *The apples* **were** *ripe*.

linking words. Words that connect and develop ideas. See **conjunction** and **conjunctive adverb**.

literal meaning. The primary, or denotative, meaning of a word. See **figurative**.

local state. The set of institutions that provide for the maintenance and protection of social, political, and economic relations at the subnational level; confusingly applied to a number of subnational forms of government, from county or municipality to province or state (here meaning a state within the U.S.).

location. Where something is. **Locations** are identified in two ways: **site** (specific location) and **situation** (relative location).

logical indicator. A word or phrase—usually a conjunction or conjunctive adverb—that shows the logical relation between sentences or clauses: e.g., *since, furthermore, therefore*.

mean, median. See **average**.

metaphor. A word picture created by replacing one element with another that is not literally applicable, but is imaginatively appropriate: e.g., *The car* **rocketed** *off the ramp and* **zoomed** *over the barrels.* (Think how much more vivid this is than *The car* **drove** *off the ramp and* **went** *over the barrels.*)

model, paradigm. Model refers to a simplified, idealized, and structured representation of a system (e.g., weather, freezing and thawing, traffic flows, or agricultural system). **Paradigm** refers to the working assumptions, procedures, and findings of scholars that collectively define particular patterns of scientific activity. Some people use the terms interchangeably, but **model** is more specific than **paradigm. Paradigm** has been used—and misused—ever since Thomas Kuhn popularized the term in his book *The Structure of Scientific Revolutions* (1962).

modifier. A word or group of words that describes or limits another element in the sentence. A **misplaced modifier** causes confusion because it is not placed next to the element it should modify: e.g., *I* **only** *ate the pie.* [Revised: *I ate only the pie.*]

mood. (a) as a grammatical term, the form that shows a verb's function (indicative, imperative, interrogative, or subjunctive); (b) when applied to literature generally, the state of mind or feeling shown.

narrative. The story or account of something, whether true or imaginary. In a novel, the narrative generally is distinguished from the dialogue.

nature. The phenomena of the physical world as a whole. A basic methodological question for geographers also is a fundamental philosophical problem: "to what extent is it proper to regard Man as a part of Nature or as standing apart from it?"[9] Another much debated issue is how "natural" the "nature" is that many environmentalists seek to preserve. These concerns reflect an awareness that nature is as much a human creation, or social entity, as a physical one.[10]

negatives. Words that are used to convey a negative meaning: e.g., *never, no, not, nothing.* Some words can be made negative by the addition of the prefixes *dis-* (dislike), *in-* (insecure), *non-* (non-appearance), *un-* (unlikely).

non-restrictive element. See **restrictive element**

noun. An inflected part of speech marking a person, place, thing, idea, action, or feeling, and usually serving as subject, object, or complement. A **common noun** is a general term: e.g., *dog, paper, automobile.* A **proper noun** is a specific name: e.g., *Mary, Ottawa, Skidoo.*

numbers, cardinal and ordinal. Cardinal numbers such as 1, 2, 3, etc., indicate how many of an item there are: e.g., *the town has 2,341 inhabitants.* **Ordinal numbers** such as 1st, 2nd, 3rd, etc., indicate rank: e.g., *the city is the 5th largest in the country.*

object. (a) a noun or pronoun that, when it completes the action of a verb, is called a **direct object**: e.g., *He passed the* **puck**. An **indirect object** is the person or thing receiving the direct object: e.g., *He passed the* **puck** (direct object) *to* **Richard** (indirect object). (b) The noun or pronoun in a group of words beginning with a preposition; pronouns take the objective case: e.g., *at the* **house**, *about* **her**, *for* **me**.

objective complement. See **complement**.

objectivity. A disinterested stance; a position taken without personal bias or prejudice. See **subjectivity**.

onomatopoeia. From the Greek for "word-making"; refers to words whose sounds echo their meaning: e.g., cats *miaow*, cows *moo*.

orient, orientate. From the Latin *oriens,* the east or sunrise, one meaning of both verbs is to place something so that it faces eastwards. More commonly, **orient** and **orientate** mean either to *adjust* or *adapt* or to *aim at* or *direct*. See also **Orientalism**.

Orientalism. In the social sciences, literary criticism, and history, **Orientalism** refers to Western "constructions" of the "East" (or, more generally, "others") and the use of Western assumptions and "findings" as the justification for control, manipulation, and incorporation. Writers such as Edward Said, Amilcar Cabral, and Joan Cocks declare that it is imperative for all subjected people to negate "Orientalism"—the colonial misrepresentations of their reality—by producing and creating their own narratives.[11]

outline. With regard to an essay or report, a brief sketch of the main parts; a written plan.

paradigm. See **model**.

paragraph. A unit of sentences arranged logically to explain or describe an idea, event, or object; usually marked by indentation of the first line.

parallel wording. Wording in which a series of items has a similar grammatical form: e.g., *At her marriage my great-grandmother promised* **to love**, **to honour**, **and to obey** *her husband*.

paraphrase. Restate in different words.

parentheses. Curved lines, enclosing and setting off a passage; not to be confused with square brackets.

parenthetical element. An interrupting word or phrase: e.g., *My musical career*, **if it can be called that**, *consisted of playing the triangle in kindergarten*.

participle. A verbal (part-verb) or verb form that functions as an adjective. Participles can be either **present**, usually marked by an -*ing* ending (e.g., *taking)*, or **past** *(having taken)*; they also can be passive *(having been taken)*.

parts of speech. The major classes of words. Some grammarians include only function words (nouns, verbs, adjectives, and adverbs); others also include pronouns, prepositions, conjunctions, and interjections.

passive voice. See **voice**.

past participle. See **participle**.

periodic sentence. A sentence in which the normal order is inverted or an essential element suspended until the very end: e.g., *Out of the house, past* *the grocery store, through the school yard, and down the railroad tracks* **raced the frightened boy**.

person. In grammar, the three classes of personal pronouns referring to the person speaking (first person), person spoken to (second person), and person spoken about (third person). With verbs, only the third person singular has a distinctive form.

personal pronoun. See **pronoun**.

personification. A stylistic technique in which objects are treated as if they were persons.

phrase. A unit of words lacking a subject-predicate combination. The most common kind is the **prepositional phrase**—a unit comprising preposition plus object. Some modern grammarians also refer to the **single-word phrase**.

place. A particular portion of the earth's surface occupied or recognized by humans who give meaning to it and derive meaning from it.

plagiarism. Using someone else's words as your own, without attribution. (See pp. 11–13).

plural. Indicating two or more in number. Nouns, pronouns, and verbs all have plural forms.

possessive case. See **case**.

précis. A brief summary of a piece of writing, usually using different words.

preface. See **foreword**.

prefix. An element placed in front of the root form of a word to make a new word: e.g., *pro-, in-, neo-, sub-*. See **suffix**.

preposition. A short word heading a unit of words containing an object, thus forming a prepositional phrase: e.g., **under** *the tree*, **before** *my time*.

process. Put most simply, the term **process** refers to the way something changes from one state to another. The physical, biological, and social sciences all study processes, of widely varying kinds.

pronoun. A word that stands in for a noun. A **personal pronoun** stands in for the name of a person: *I, he, she, we, they*, etc.

proofreading. The critically important process of checking written work (whether an essay, a lab report, or whatever else) for mistakes of any kind before handing it in.

punctuation. A conventional system of signs used to indicate stops or divisions in a sentence and to make meaning clearer: e.g., comma, period, semicolon, etc. (see Chapter 16).

qualifier. A word that intensifies or reinforces other words: e.g., *These soils are* **very** *good for potatoes.*

quotation. The exact replication of someone else's words. See **quotation marks** in Chapter 11 for when and when not to use quotation marks.

race, racial, racialism, racism, racist. A **race** is a group of persons or animals or plants connected by common ancestry and classified on the basis for genetically derived physical characteristics. The adjective **racial** can refer either to race itself ("*racial characteristics*") or to difference in race ("*racial discrimination*"). **Racialism** and **racism** refer to the belief in the superiority of a particular race, and to behaviours that express antagonism between people of different races, hence **racist** attitudes and behaviour. See also **culture** and Chapter 13.

redundancy. The use of more words than are necessary: e.g., *The* **three triplets** *are all in my class.* See **tautology**.

reference works. Material consulted when preparing an essay or report.

referent (antecedent). The noun for which a pronoun stands.

reflexive pronoun. A pronoun that ends in *-self* or *-selves* and so refers to the subject: e.g., *The professor cut* **himself** *on the beaker.* Don't make the mistake of saying things like, "Send the material to myself" (when you should say ". . . to me").

reflexive verb. A verb that has the same person as subject and object: e.g., *The student taught himself how to use the computer.*

region, regionalism. Region refers to an area defined according to a particular set of criteria; all regions are human creations in the sense that the criteria people select determine the areal extent of the region. A distinction is made between formal and functional regions. Formal regions are defined according to distribution (e.g., of loess soil in Europe, or milk production in Wisconsin), while functional regions are defined according to the **areal** extent of the linkages within a system (e.g., a settlement system, with reference to the area that is functionally linked to a large central place—such as Seattle's "urban field"). The concept of region is used in geography and in urban and regional planning. **Regionalism** reflects a sense of togetherness among the people of a relatively large area. It is recognizable only in the context of that area's being only a part of an even larger area (for instance, "southern" regionalism within the U.S.). When a regionalism finds political expression within the larger state and its leaders give primacy to the interests of their region, thus undercutting the role and, perhaps, the legitimacy of the state, the result is **sectionalism**. **Sectionalism** may lead to *self-determination* and *secession*.[12] Thus sectionalism operates as a centrifugal force within a state. (See also **classification** and **territory**.)

register. The degree of formality in word choice and sentence structure.

relative clause. A clause headed by a relative pronoun: e.g., *the man* **who came to dinner** *is my uncle*.

relative pronoun. *Who, which, what, that,* or their compounds beginning an adjective or noun clause: e.g., *the house* **that** *Jack built*; **whatever** *you say*.

restrictive element. A phrase or clause that identifies or is essential to the meaning of a term: e.g., *The book* **that I need** *is lost*. It should not be set off by commas. A non-restrictive element is not needed to identify the term and is usually set off by commas: e.g., *This book,* **which I got from my aunt,** *is one of my favourites*.

restructuring. The wide range of fundamental economic and social changes made as national societies seek to come to terms with shifts in the global economy.[13] See **globalization**.

rhetoric/oratory. Language that is intended to persuade listeners; **oratory** refers only to spoken language.

rhetorical questions. Questions posed and answered by a speaker (or writer) for stylistic effect, to draw attention to a point; no audience response is needed or expected: e.g., *"What am I to you?" asked the Minister of the Environment. "I am the person in government who cares most about environmental issues."*

root. The part of a word that carries its essential meaning: e.g., *spac-* is the root of *spacious* because it carries the meaning "space."

run-on sentence. A sentence that goes on beyond the point where it should have stopped. The term covers both the **comma splice** (two sentences joined by a comma) and the **fused sentence** (two sentences joined without any punctuation between them).

scale. The level of representation of reality; an important concept in cartography. Confusingly, large-scale maps depict small areas (but can include considerable detail), whereas small-scale maps depict large areas (and thus include less detail). In geography, the scale selected for any study will help to determine the degree of focus and, thus, the conclusions that can be drawn from the analysis.

sectionalism. See **region.**

sentence. A grammatical unit that includes both a subject and a predicate. The end of a sentence is marked by a period.

sentence fragment. A group of words lacking either a subject or a verb; an incomplete sentence.

sewage, sewerage. Sewage is waste matter which goes down the sewer. **Sewerage** refers to the process whereby waste material or sewage is removed by means of sewers; also, it may mean a system of sewers.

sexist language. Language that excludes or demeans one sex or the other. See Chapter 13.

simile. A figure of speech in which one thing is explicitly likened to something else: e.g., *The lawn is* **as** *smooth* **as** *a billiard table*; *The lake is* **like** *glass*. Similes are neither as direct nor as vivid as **metaphors.**

simple sentence. A sentence made up of only one clause: e.g., *Joan climbed the tree.*

site. The *specific* aspect of **location.** A specific site cannot change, although its internal characteristics can be altered. See **situation.**

situation. The *relative* aspect of **location.** The relative location of a site will change if alterations are made to external linkages. See **site.**

slang. Colloquial speech, inappropriate for academic writing; often used in a special sense by a particular group: e.g., *gross* for *disgusting*; *gig* as a musician's term.

space. The continual expanse in which things exist and move; a portion of that continual expanse; the interval between locations. Space is a central concept for geographers, who are concerned with the absolute, relative, and relational nature of space over the surface of the earth. See **location**, **place**, **region**, **spatial**, and **territory**.

spatial. The adjectival form of **space**, as in *spatial analysis*, *spatial structure*, *spatial interaction*, *spatial preference*. Note that it is spelled with a *t*, not a c. **Spatial interaction** is the movement of goods, people, and ideas between interdependent regions. See **complementarity**.

split infinitive. A construction in which a word is placed between *to* and the base verb: e.g., *to completely finish*. Many people object to this kind of construction, but splitting infinitives is sometimes necessary when the alternatives are awkward or ambiguous.

squinting modifier. A kind of misplaced modifier; one that could be connected to elements on either side, making meaning ambiguous: e.g., *When he wrote the letter* **finally** *his boss thanked him*.

stalactite, stalagmite. Both are deposits formed by dripping water in limestone caves: a **stalactite** hangs like an icicle from the roof of a cave; a **stalagmite** builds up from the floor.

standard English. The English spoken and written by literate people; or, the set of words, grammar, and basic linguistic sounds that are used by them. There are no set rules for governing what is accepted as the English language develops. (In contrast, the *Académie française* decides what, and what not, to accept as appropriate French.) Two broadly-based forms of English are recognized: British-English and American-English. British-English is sometimes referred to as International English, since it extends from Britain throughout the Commonwealth, and more widely. "Non-standard" English (some of it class-based) can be found in numerous locales as well as broad regional settings (with some unique words) within Britain and the U.S., and also in Canada, New Zealand, Australia, South Africa, India, the West Indies, and so on. In addition, of course, there are different accents everywhere!

stem. The basic word to which prefixes and suffixes are added. The stem remains unchanged: e.g., *agriculture*, *agricultural*, *agriculturist*, *monoagriculture*.

subject. In grammar, the noun or noun equivalent about which something is predicated; that part of a clause with which the verb agrees: e.g., **They** *swim every day when the* **pool** *is open.*

subjective complement. See **complement**.

subjectivity. A personal stance, not impartial. See **objectivity**.

subordinate clause. See **clause**.

subordinating conjunction. See **conjunction**.

subordination. Making one clause in a sentence dependent on another.

suffix. An addition placed at the end of a word to form a derivative: e.g., *prepare—pre*pare**tion**; *sing—sing***ing**. See **prefix**.

symbol (in science). Shortened form of a term: e.g., CO_2 stands for *carbon dioxide*; *Hz* stands for *hertz*. See also **symbolism**.

symbolism. The use of one thing to represent or suggest something other than itself: e.g., white (in some cultures) is a symbol of innocence; the maple leaf is a symbol of Canada.

synonyms. Words with the same dictionary meaning: e.g., *begin* and *commence*.

syntax. Sentence construction; the grammatical relations of words.

tautology. Repetition using different words, usually unnecessarily: e.g., *The tractor* **returned back** *to the shed.*

technical writing. Writing by a specialist for other specialists—as in soils engineering or fluvial geography—using technical words and phrases that may not be familiar to outside readers.

tense. The time reference of verbs.

territoriality. The attempt by an individual or group to influence or establish control over a particular **territory**. See **region, regionalism**, and **sectionalism**.

territory. A particular portion of the surface of the earth. The term implies bounding, either formally or informally, by people who either hold or covet the area. Territory is both a physical and a psychological phenomenon, for people "see" meaning in and "obtain" meaning from it, perceiving the territory and its landscape as living entities that are filled with meaning. Such perceptions reflect cultural beliefs that exist only within the "geographies of the mind"; yet any group's **cultural ecology** and spatial patterning are powerfully influenced by their beliefs. The link between a people and "their" territory is complex. Having a territory offers a people security (to look inward to preserve the integrity of the society against outside forces) and opportunity (from which to reach out to other societies).[14] See **boundary**, **cultural landscape**, **space**, and **territoriality**.

theme. A recurring or dominant idea.

thesis statement. A one-sentence assertion that gives the central argument of an essay or thesis.

topic sentence. The sentence in a paragraph (usually at the beginning) that expresses the main or controlling idea.

transition word. A word that shows the logical relation between sentences or parts of a sentence and thus helps to signal the change from one idea to another: e.g., *therefore*, *also*, *accordingly*.

transitive verb. One that takes an object: e.g., *hit*, *bring*, *cover*.

usage. Accepted practice.

verb. That part of a predicate expressing an action, state of being, or condition, telling what a subject is or does. Verbs inflect to show tense (time). The principal parts of a verb are the three basic forms from which all tenses are made: the base infinitive, the past tense, and the past participle.

verbal. A word that is similar in form to a verb but does not function as one: a participle, a gerund, or an infinitive.

voice. The form of a verb that shows whether the subject acted (active voice) or was acted upon (passive voice): e.g., *He* **hit** *the ball* (active). *The ball* **was hit** *by him* (passive). Only transitive verbs (verbs taking objects) can be passive.

NOTES

1   Anssi Paasi, *Territories, Boundaries and Consciousness* (Chichester: Wiley, 1996).
2   Yi-Fu Tuan, "A View of Geography," *Geographical Review*, Vol. 81 (1991), p. 99; the full article is on pp. 99–107. See also Geoffrey J. Martin, *All Possible Worlds:*

A History of Geographical Ideas, 4th ed. (New York: Oxford University Press, 2005).

3   John E. Chappell, Jr., "Relations Between Geography and Other Disciplines," in *On Becoming a Professional Geographer*, M.S. Kenzer, ed. (Columbus: Merrill, 1989), pp. 17–31.

4   Anne Buttimer, *The Practice of Geography* (London: Methuen), 1983, p. 12.

5   For discussions of the full array of specialities, methods, techniques, and applications see the references listed in endnote 1 in Chapter 2.

6   Harm J. de Blij, *Why Geography Matters: Three Challenges Facing America* (New York: Oxford University Press, 2005).

7   Denis Cosgrove and Stephen Daniels, *The Iconography of Landscape* (Cambridge: Cambridge University Press, 1998); Peter Atkins, I. Simmons, and B. Roberts, *People, Land and Time: An Historical Introduction to the Relations Between Landscape, Culture and Environment* (London: Arnold, 1998); and Brett Wallach, *Understanding the Cultural Landscape* (New York: Guilford, 2005).

8   David Lowenthal, *The Heritage Crusade and the Spoils of History* (Cambridge: Cambridge University Press, 1998).

9   R.J. Chorley, "Geography as Human Ecology," in *Directions in Geography*, R.J. Chorley, ed. (London: Methuen; New York: Barnes and Noble, 1973), p. 158.

10 Neil Evernden, *The Social Creation of Nature* (Baltimore: Johns Hopkins University Press, 1992).

11 Edward W. Said, *Orientalism* (New York: Vintage Books, 1979), and Anne Godlewska and Neil Smith, eds., *Geography and Empire* (Oxford: Blackwell, 1994).

12 David B. Knight, "People Together, Yet Apart: Rethinking Territory, Sovereignty, and Identities," in *Reordering the World: Geopolitical Perspectives on the 21st Century*, 2nd ed., George J. Demko and William B. Wood, eds. (Boulder, CO: Westview, 1999), pp. 209–26.

13 David B. Knight and Alun E. Joseph, *Restructuring Societies: Insights from the Social Sciences* (Ottawa: Carleton University Press, 1999).

14 Jean Gottmann, *The Significance of Territory* (Charlottesville: University Press of Virginia, 1973), and Robert David Sack, *Human Territoriality: Its Theory and History* (Cambridge: Cambridge University Press, 1986).

Appendix I

Weights, Measures, and Notation

The conversion factors are not exact unless so marked. They are given only to the accuracy likely to be needed in everyday calculations.

1. IMPERIAL AND AMERICAN, WITH METRIC EQUIVALENTS

Linear measure

1 inch	= 25.4 millimetres exactly
1 foot = 12 inches	= 0.3048 metre exactly
1 yard = 3 feet	= 0.9144 metre exactly
1 (statute) mile = 1,760 yards	= 1.609 kilometres
1 int. nautical mile	
= 1.150779 miles	= 1.852 km exactly

Square measure

1 square inch	= 6.45 sq. centimetres
1 square foot = 144 sq. in.	= 9.29 sq. decimetres
1 square yard = 9 sq. ft.	= 0.836 sq. metre
1 acre = 4,840 sq. yd.	= 0.405 hectare
1 square mile = 640 acres	= 259 hectares

Cubic measure

1 cubic inch	= 16.4 cu. centimetres
1 cubic foot = 1,728 cu. in.	= 0.0283 cu. metre
1 cubic yard = 27 cu. ft.	= 0.765 cu. metre

Capacity measure

Name	System	Equal to	Metric
fluid oz.	imperial	1/20 imp. pint	28.41 ml
	US (liquid)	1/16 US pint	29.57 ml
gill	imperial	1/4 pint	142.07 ml
	US (liquid)	1/4 pint	118.29 ml
pint	imperial	20 fl.oz.(imp.)	568.26 ml
	US (liquid)	16 fl.oz.(US)	473.18 ml
	US (dry)	1/2 quart	550.61 ml
quart	imperial	2 pints	1.1365 litres
	US (liquid)	2 pints	0.9464 litre
	US (dry)	2 pints	1.1012 litres
gallon	imperial	4 quarts	4.546 litres
	US (liquid)	4 quarts	3.785 litres
peck	imperial	2 gallons	9.092 litres
	US (dry)	8 quarts	8.810 litres
bushel	imperial	4 pecks	36.369 litres
	US (dry)	4 pecks	35.239 litres

Avoirdupois weight

1 grain	= 0.065 gram
1 dram	= 1.772 grams
1 ounce = 16 drams	= 28.35 grams
1 pound = 16 ounces	
= 7,000 grains	= 0.45359237
	kilogram exactly
1 stone = 14 pounds	= 6.35 kilograms
1 quarter = 2 stones	= 12.70 kilograms
1 hundredweight = 4 quarters	
= 112 lb.	= 50.80 kilograms
1 (long) ton = 20 cwt. = 2,240 lb.	= 1.016 tonnes
1 short ton = 2,000 pounds	= 0.907 tonne

2. METRIC, WITH IMPERIAL EQUIVALENTS

Linear measure

1 millimetre	= 0.039 inch
1 centimetre = 10 mm	= 0.394 inch
1 decimetre = 10 cm	= 3.94 inches
1 metre = 100 cm	= 1.094 yards
1 decametre = 10 m	= 10.94 yards
1 hectometre = 100 m	= 109.4 yards
1 kilometre = 1,000 m	= 0.6214 mile

Square measure

1 square centimetre	= 0.155 sq. inch
1 square metre = 10,000 sq. cm	= 1.196 sq. yards
1 are = 100 sq. metres	= 119.6 sq. yards
1 hectare = 100 ares	= 2.471 acres
1 square kilometre = 100 ha	= 0.386 sq. mile

Cubic measure

1 cubic centimetre	= 0.061 cu. inch
1 cubic metre = one million cu. cm	= 1.308 cu. yards

Capacity measure

1 millilitre	= 0.002 pint (imperial)
1 centilitre = 10 ml	= 0.018 pint
1 decilitre = 100 ml	= 0.176 pint
1 litre = 1,000 ml	= 1.76 pints
1 decalitre = 10 l	= 2.20 gallons (imperial)
1 hectolitre = 100 l	= 2.75 bushels (imperial)

Weight

1 milligram	= 0.015 grain
1 centigram = 10 mg	= 0.154 grain
1 decigram = 100 mg	= 1.543 grain
1 gram = 1,000 mg	= 15.43 grain
1 decagram = 10 g	= 5.64 drams
1 hectogram = 100 g	= 3.527 ounces
1 kilogram = 1,000 g	= 2.205 pounds
1 tonne (metric ton) = 1,000 kg	= 0.984 (long) ton

3. SI UNITS

Base units

Physical quantity	Name	Abbr. or symbol
length	metre	m
mass	kilogram	kg
time	second	s
electric current	ampere	A
temperature	kelvin	K
amount of substance	mole	mol
luminous intensity	candela	cd

Supplementary units

Physical quantity	Name	Abbr. or symbol
plane angle	radian	rad
solid angle	steradian	sr

Derived units with special names

Physical quantity	Name	Abbr. or symbol
frequency	hertz	Hz
energy	joule	J
force	newton	N
power	watt	W
pressure	pascal	Pa
electric charge	coulomb	C
electromotive force	volt	V
electric resistance	ohm	Ω
electric conductance	siemens	S
electric capacitance	farad	F
magnetic flux	weber	Wb
inductance	henry	H
magnetic flux density	tesla	T
luminous flux	lumen	lm
illumination	lux	lx

4. TEMPERATURE

Celsius (or Centigrade): Water boils (under standard conditions) at 100° and freezes at 0°

Fahrenheit: Water boils at 212° and freezes at 32°

Kelvin: Water boils at 373.15 kelvins and freezes at 273.15 kelvins.

Celsius	Fahrenheit
-17.8°	0°
-10°	14°
0°	32°
10°	50°
20°	68°
30°	86°
40°	104°
50°	122°
60°	140°
70°	158°
80°	176°
90°	194°
100°	212°

To convert Celsius into Fahrenheit: multiply by 9, divide by 5, and add 32.

To convert Fahrenheit to Celsius: subtract 32, multiply by 5, and divide by 9.

5. METRIC PREFIXES

	Abbr. or symbol	Factor
deca-	da	10
hecto-	h	10^2
kilo-	k	10^3
mega-	M	10^6
giga-	G	10^9
tera-	T	10^{12}
peta-	P	10^{15}
exa-	E	10^{18}
deci-	d	10^{-1}
centi-	c	10^{-2}
milli-	m	10^{-3}
micro-	μ	10^{-6}
nano-	n	10^{-9}
pico-	p	10^{-12}
femto-	f	10^{-15}
atto-	a	10^{-18}

These prefixes may be applied to any units of the metric system: hectogram (abbr. hg) = 100 grams; kilowatt (abbr. kW) = 1,000 watts; megahertz (MHz) = 1 million hertz; centimetre (cm) = $^1/_{100}$ metre; microvolt (μV) = one millionth of a volt; picofarad (pF) = 10^{-12} farad, and are sometimes applied to other units (megabit).

6. POWER NOTATION

This expresses concisely any power of ten (any number that is composed of factors of 10). 10^2 or ten squared = $10 \times 10 = 100$; 10^3 or ten cubed = $10 \times 10 \times 10 = 1,000$. Similarly, $10^4 = 10,000$ and $10^{10} = 1$ followed by ten zeros = 10,000,000,000. Proceeding in the opposite direction, dividing by ten and subtracting one from the index, we have $10^2 = 100$, $10^1 = 10$, $10^0 = 1$, $10^{-1} = ^1/_{10}$, $10^{-2} = ^1/_{100}$, and so on; $10^{-10} = ^1/_{10,000,000,000}$.

7. BINARY SYSTEM

Only two units (0 and 1) are used, and the position of each unit indicates a power of two.

One to ten written in binary form:

	eights (2^3)	fours (2^2)	twos (2^1)	one
1				1
2			1	0
3			1	1
4	1	0	0	0
5	1	0	0	1
6	1	0	1	0
7	1	0	1	1
8	1	0	0	0
9	1	0	0	1
10	1	0	1	0

I.e., ten is written as 1010 ($2^3 + 0 + 2^1 + 0$); one hundred is written as 1100100 ($2^6 + 2^5 + 0 + 0 + 2^2 + 0 + 0$).

Appendix II

Selected Journals
of Interest to Geographers and
Environmental Scientists

GENERAL JOURNALS

Annales de géographie

Annals of the Association of American
   Geographers

Area

Environmental Management

Erdkunde

Focus

Geoforum

Geographical Journal

Geographical Review

Geographische Zeitschrift

Geography Research Forum

GeoJournal

International Journal of Environmental
   Studies

Journal of Geography

Landscape

Nature

New Scientist

Progress in Development Studies

Progress in Environmental Science

Progress in Human Geography

Progress in Physical Geography

Science

Scientific American

The Professional Geographer

Tijdschrift voor economische en sociale
   geografie

Transactions of the Institute of British
   Geographers

SPECIALIST THEMATIC JOURNALS

Antipode

Applied Geography

Canadian Journal of Earth Sciences

Cartographica: The International Journal
   for Geographic Information and
   Geovisualization

Climate Change

Cultural Geographies

Earth Surface Processes and Landforms

Ecology

Economic Geography

Environment and Behavior

Environment and Planning A

Environment and Planning B–Planning
   and Design

Environment and Planning C–Government
   and Policy

Environment and Planning D–Society and
   Space

Ethics, Place and Environment

Ethnic and Racial Studies

Gender & Development

Gender, Place and Culture: A Journal of
Feminist Geography
Geografiska Annaler Series A–Physical
Geography
Geografiska Annaler Series B–Human
Geography
Geographical Analysis: An International
Journal of Theoretical Geography
Geographical Research
Geographic Information Sciences
Géographie et cultures
Géographie physique et quaternaire
Geomorphology
Geopolitics
GIScience and Remote Sensing
Global Ecology and Biogeography
Global Networks
GPS World
Health and Place
International Journal of Fieldwork Studies
International Journal of Remote Sensing
International Migration
Journal of Biogeography
Journal of Cultural Geography
Journal of Economic Geography
Journal of Environmental Management
Journal of Ethnic and Migration Studies
Journal of Historical Geography

Journal of Political Ecology
Journal of Rural Studies
Journal of Sustainable Tourism
Journal of Transportation and Statistics
L'Espace géographique
L'Espace politique (online)
Mobilities
National Identities
Natural Resources Journal
Photogrammetric Engineering and Remote
Sensing
Physical Geography
Political Geography
Population and Development Studies
Population, Space and Place
Regional Studies
Social and Cultural Geography
Sport Place: An International Journal of
Sports Geography
The Cartographic Journal
Tourism Geographies: International
Journal of Tourism Space, Place and
Environment
Tourism in Marine Environments
Tourism Recreation Research
Urbana
Urban Geography
Urban Studies

REGIONAL PROFESSIONAL JOURNALS

Acta Geographica
Ambio
Arctic
Asia Pacific Viewpoint
Australian Geographer
Cahiers de géographie du québec
Canadian Geographer/Le Géographe
Canadien

Cybergeo: European Journal of Geography
Etudes Canadiennes/Canadian Studies
Eurasian Geography and Economics
Geoforum
Geographical Research (formerly
Australian Geographical Studies)
Irish Geography
Journal of Central Asian Studies

Journal of Latin American Geography
New Zealand Geographer
Norwegian Journal of Geography
Polar Geography
Polar Record
Scottish Geographical Journal (formerly
  Scottish Geographical Magazine)
Singapore Journal of Tropical Geography

South African Geographical Journal
Swansea Geographer
The Arab World Geographer
The Great Lakes Geographer
The North American Geographer
The Southeastern Geographer
Transactions of the Institute of British
  Geographers

PEDAGOGICAL JOURNALS

Geography
Journal of Geography
Journal of Geography in Higher Education
New Zealand Journal of Geography

The Monograph (Ontario Association
  for Geographic and Environmental
  Education)

POPULAR JOURNALS FOR THE GENERAL PUBLIC

American Geographical Society's
  FOCUS on Geography
Australian Geographic
Canadian Geographic

Geographical Magazine
National Geographic
New Zealand Geographic

ONLINE JOURNALS

ACME: An International E-Journal for
  Critical Geographies
African Studies Quarterly
Contemporary Disaster Review
Geobase
Geography Compass
Geography Online
Geomaya
Geoweb Online

Global Networks
International Journal of Health
  Geographics
Journal of Geographic Information and
  Decision Analysis
Journal of Hospitality, Leisure, Sport and
  Tourism Education
Journal of Political Ecology
Mappa.Mundi Magazine

Appendix III

Selected Websites of Interest to Geographers and Environmental Scientists

The following (selective) URL list includes websites of special interest to geographers and others concerned with environmental issues. Some offer access to specific agencies and their publications, and many offer links to other useful sites.

About.com Guide to Geography: http://geography.about.com
Agricultural and Related Information Site Index:
    http://agrigator.ifas.ufl.edu/ag.htm
Agriculture and Agri-Food Canada: http://www.agr.gc.ca
American Geographical Society: http://www.amergeog.org
Association of American Geographers (AAG): http://www.aag.org
Australian Environment: http://environment.gov.au
BBC Online: http://www.bbc.co.uk
Bioline International open access journals: http://www.bioline.org.br
Biology Browser: http://www.biologybrowser.com
Botany, Internet Directory: http://www.botany.net/IDB/botany.html
Canada – Department of Fisheries and Oceans: http://www.ncr.dfo.ca
Canada – Environment Canada: http://www.ec.gc.ca
Canada – Geological Survey: http://gsc.nrcan.gc.ca
Canada – Government of Canada internet addresses:
    http://canada.gc.ca/directories/internet_e.html
Canada – National Atlas: http://www.atlas.gc.ca
Canada – National Library and Archives: http://www.nlc-bnc.ca
Canada – Natural Resources: http://www.nrcan.gc.ca
*Canada – Office of the Auditor General – reports of the Commissioner of
    Environment and Sustainable Development*: http://www.oag-bvg.gc.ca/
    internet/English/cesd_fs_e_921.html
Canada – Parks Canada Agency: http://www.pc.gc.ca

Canada – Statistics Canada: http://www.statcan.ca

Canadian Association of Geographers (CAG): http://www.cag_acg.ca/en

Canadian Broadcasting Corporation (CBC): http://www.cbc.ca

Canadian Council for Geographic Education: http://ccge.org/ccge

Canadian Environmental Assessment Agency: http://www.ceaa.gc.ca

Canadian Legal Information Institute: http://www.canlii.org

Canadian Society of Soil Science: http://www.csss.ca/

Canadian Water Resources Association: http://www.cwra.org/

Canadian Wildlife Service: http://www.cws-scf.ec.gc.ca

Center for International Earth Science Information Network (CIESIN):
      http://www.ciesin.org

Climate Change – Intergovernmental Panel on Climate Change:
      http://www.ipcc.ch (see *Presentations & Graphics*)

Committee on the Status of Endangered Wildlife in Canada:
      http://www.cosewic.gc.ca

*Demography – U.S. and International (see also U.N. and Statistics
      Canada)*: http://www.census.gov *and* http://census.gov/ipc/www/

Directory of open access journals: http://www.doaj.org

Earth Observing System: http://eospso.gsfc.nasa.gov

Earth Science Data Directory: http://www.dmoz.org/science

Earthweek (weekly summaries of natural and human-made events):
      http://www.earthweek.com

East and Southeast Asia – Annotated Directory: http://newton.uor.edu/
      Departments&Programs/AsianStudiesDept/index.html

Encyclopaedia Britannica Online: http://www.britannica.com

EnviroLink: http://www.envirolink.org

Environmental Activism (incl. international):
      http://www.ecomall.com/activism

Federation of Canadian Municipalities: http://www.fcm.ca/

Fisheries and Oceans Canada: http://www.dfo-mpo.gc.ca

Geographical Association (U.K.): http://www.geography.org.edu

Geographical Learning for Sustainable Development:
      http://www.aag.org/sustainable

Geographical Resources: http://www.library.uu.nl/geosource *and*
      http://www.library.wisc.edu/libraries/Geography/Geo_Resources/
      index.html

*Geographical Resources: American Geographical Society Library ("Current
      Geographical Publications" and "GeoBib Search")*:
      http://www.uwm.edu/Library/AGSL

Geography Departments Worldwide: http://univ.cc/geolinks *and*
http://www.igu_net.org/uk/igubase/geography_departments.html
Geography Network: http://www.geographynetwork.com
GeoHive: Selected Global Statistics: http://www.xist.org
*Geo-Images Project (copyrighted slides on geographic topics; permission
to use some of them can be sought by e-mail)*:
http://geogweb.berkeley.edu/GeoImages.html
Geospatial Datasets: http://www.colorado.edu/geography/virtdept/
resources/data/data.htm
Global Change Master Directory: http://gcmd.nasa.gov
*Global Change Newsletter (International Geosphere-Biosphere
Programme)*: http://www.igbp.net
Google geography: http://www.google.com/search?q=geography
Great Globe Gallery, The: http://www.staff.amu.edu.pl/~zbzw/glob/
glob1.htm
Home of Geography: http://www.homeofgeography.org
Human Rights Watch: http://www.hrw.org
Indian and Northern Affairs Canada:
http://www.ainc-inac.gc.ca/index-eng.asp
Institute of British Geographers (IBG): http://www.rgs.org
International Council for Science (ICSU): http://www.icsu.org/index.php
International Geography Olympiad (IGEO): http://www.geoolympiad.org
International Governmental Organizations and Agencies: http://www.lib.
umich.edu/govdocs/intl.html *and* http://www.library.northwestern.
edu/govpub/resource/internat/igo.html
International Joint Commission: http://www.ijc.org
International Labour Organization: http://www.ilo.org
Internet Public Library: http://www.ipl.org
Librarians' Internet Index (see, e.g., Science): http://www.lii.org
NASA: http://www.nasa.gov
NASA Global Change Master Directory: http://gcmd.gsfc.nasa.gov
National Council for Geographic Education (U.S.): http://www.ncge.org
National Energy Board: http://www.neb.gc.ca
National Geographic Society (U.S.): http://www.nationalgeographic.com
*National Geographic Society – maps and geography (incl. free outline
maps)*: http://www.nationalgeographic.com/resources/ngo/maps
and http://www.nationalgeographic.com/xpeditions
National Round Table on the Environment and the Economy (NRTEE):
http://www.nrtee-trnee.ca/

National Society of Consulting Soil Scientists, Inc. (U.S.A.):
http://www.nscss.org

NationMaster.com (a massive central data source; you can graphically compare nations using map and graphs):
http://www.nationmaster.com/index.php

Natural Resources Canada: http://www.nrcan.gc.ca

Newspapers: http://www.onlinenewspapers.com *and* http://www.newspapers.com *and* http://www.world_newspapers.com/index.html

New Zealand – Environment: http://www.mfe.govt.nz

New Zealand – Environmental and Conservation Organizations:
http://www.eco.org.nz/links.asp

NGO links: http://www.ngo.org/links/index.htm

Oddens Bookmarks (maps and mapping):
http://oddens.geog.uu.nl/index.php

Ontario Association for Geographic and Environmental Education:
http://www.oagee.org

Resources for Geographers: http://www.colorado.edu/geography/virtdept/resources/contents.htm

Royal Canadian Geographical Society: http://www.rcgs.org

Royal Geographical Society, (U.K.): http://www.rgs.org

Scholarly Societies Project (links to about 3,800 scholarly societies):
http://www.scholarly-societies.org

Science Conservation Council of Canada: http://www.soilcc.ca

Smithsonian (U.S.): http://www.si.edu

Smithsonian Museum of Natural History: http://www.mnh.si.edu

Social Science Information Gateway: Environmental Sciences:
http://www.sosig.ac.uk/environmental_sciences_and_issues

Social Science Information Gateway: Geography:
http://www.sosig.ac.uk/geography

Social Science Information Gateway: GIS and Cartography:
http://www.sosig.ac.uk/roads/subject_listing/World_cat/gis.html

Sustainable Development: http://www.ulb.ac.be/ceese/meta/sustvl.html

U.K. – British Library: http://www.bl.uk

U.K. Stationery Office: http://www.tsoshop.co.uk

United Nations (U.N.): http://www.un.org

U.N. Cartographic Section:
http://www.un.org/Depts/Cartographic/english/htmain.htm

U.N. Department of Economic and Social Affairs, Division for Sustainable Development: http://www.un.org/esa/sustdev/

U.N. Development Programme: http://www.undp.org

U.N. Environment Network: http://www.unep.net

U.N. Framework Convention on Climate Change: http://unfccc.int/2860.php

U.N. Refugee Agency: http://www.unhcr.org

UNESCO sources and links: http://portal.unesco.org/ci/en

UNESCO World Heritage List: http://whc.unesco.org

UNICEF (for country stats): http://www.unicef.org

Union of Concerned Scientists: http://www.ucsusa.org

U.S. Census, numerous sites including: http://www.census.gov *and* http://www.census.gov/geo/www/maps *and* http://factfinder.census.gov *and* http://www.census.gov/geo/www/index.html

U.S. CIA ("World Handbook"): http://www.odci.gov/cia/publications/factbook/index.html

U.S. Environmental Protection Agency: http://www.epa.gov

U.S. Geological Survey, numerous sites, including: http://www.usgs.gov *and* http://www.usgc.gov/science *and* http://nationalmap.gov *and* http://geography.usgs.gov

U.S. Geospatial One-Stop: http://www.geodata.gov

U.S. Government Printing Office: http://www.access.gpo.gov

USGS Earthshots: Satellite Images of Environmental Change: http://edcwww.cr.usgs.gov/earthshots/slow/tableofcontents

U.S. Library of Congress: http://www.loc.gov/index.html

U.S. Library of Congress, Geography and Map Reading Room: http://www.loc.gov/rr/geogmap

U.S. National Atlas: http://www.nationalatlas.gov

U.S. State and Local Government on the Net: http://www.statelocalgov.net

U.S. State Department ('Background Notes'): http://www.state.gov/r/pa/ei/bgn

World Bank: http://www.worldbank.org

World Health Organization: http://www.who.int/en/

World Meteorological Organization: http://www.wmo.int/pages/index_en.html

World Trade Organization: http://www.wto.org

Worldwatch Institute: http://www.worldwatch.org

WWW Resources in Economics: http://www.helsinki.fi/WebEc
WWW Virtual Library, The: http://www.vlib.org
Yahoo geography: http://www.yahoo.com/Science/Geography

Index